Resurrecting Extinct Species

Douglas Ian Campbell
Patrick Michael Whittle

Resurrecting Extinct Species

Ethics and Authenticity

Douglas Ian Campbell
Philosophy Department
University of Canterbury
Christchurch, New Zealand

Patrick Michael Whittle
Christchurch, New Zealand

ISBN 978-3-319-69577-8   ISBN 978-3-319-69578-5   (eBook)
https://doi.org/10.1007/978-3-319-69578-5

Library of Congress Control Number: 2017955701

Cover illustration: Cover pattern © Melisa Hasan

Printed on acid-free paper

This Palgrave Macmillan imprint is published by Springer Nature
The registered company is Springer International Publishing AG
The registered company address is: Gewerbestrasse 11, 6330 Cham, Switzerland

Dedication
From Doug—for Zoe, Isla, Nelly and Lucia.
From Mick—for Sue and Poppy.

PREFACE

Momentous things are afoot in laboratories around the world. DNA, extracted from the bones of extinct animals, is being put into sequencing machines, and the long-lost genetic blueprints of extinct species are being pieced together by computers. Synthetic copies of extinct animals' genes are being inserted into the chromosomes of living cells. Such modified cells are being turned into full grown, transgenic organisms by cloning. Doctor Frankenstein himself would be impressed.

The technology for bringing back extinct species is now a reality. It is growing more powerful day by day, for in the field of synthetic biology, what was science fiction one year is often humdrum, routine procedure only a few years later. Of course, the technology has certain fundamental limits. The *Tyrannosaurus rex* of *Jurassic Park* fame will never tread the Earth again, any remaining traces of its DNA having been bombarded to smithereens by cosmic radiation long ago. 'All the king's horses and all the king's men will not put the T. rex together again'—a fact that will be a matter of eternal regret to generations of schoolchildren down the centuries to come. But the news for schoolchildren is not all bad. DNA's half-life in permafrost is very long, which puts the woolly mammoth in firm contention as a de-extinction candidate.

From a philosopher's perspective, de-extinction is a delightfully controversial subject. The questions it raises are immense, important, and difficult. They go to the heart of much more general philosophical problems. What is a species? Why should we value biodiversity? Do we have a moral duty to undo harm that our species has done to other species? Does moral wisdom militate against our using technology to manage and control the non-human world?

Given the pace of technological progress, a book on the philosophy of de-extinction is urgently needed. And here it is! We hope the book will be useful. We believe it contains some good ideas. We are acutely conscious that it is not perfect. We hope we have not said too many unwise things, or criticized others views too unkindly or unfairly.

We would like to thank Zoe Reeves, Michael-John Turp, Carolyn Mason and Hazel and David Conroy for their many helpful suggestions.

CONTENTS

CHAPTER 1

Conservation in a Brave New World

Abstract This chapter introduces the two main philosophical questions that are raised by the prospect of extinct species being brought back from the dead—namely, the 'Authenticity Question' and the 'Ethical Question'. It distinguishes different types of de-extinction, and different methods by which de-extinction can be accomplished. Finally, it examines the aims of wildlife conservation with a view to whether they are compatible with de-extinction, or not.

Keywords De-extinction • Conservation • Authenticity • Ethics

1.1   THE LAST BUCARDO (OR NOT?)

In January 2000, a storm in Spain's Ordesa National Park caused a tree bough to snap and fall. Sheltering beneath the tree was an animal named 'Celia', the last Pyrenean ibex, or 'bucardo', left on Earth. Mortally injured, Celia breathed her last breath—and with that the goat-like bucardo, long persecuted by hunters, was extinct.

This, however, was not to be the last breath of an animal with bucardo DNA.

Months before Celia died, a sample of skin had been snipped from her ear and preserved in liquid nitrogen. In October 2000, a team of reproductive physiologists set to work on Celia's cryogenically preserved cells, applying the techniques that had been used in 1996 to clone 'Dolly', the

© The Author(s) 2017
D.I. Campbell, P.M. Whittle, *Resurrecting Extinct Species*,
https://doi.org/10.1007/978-3-319-69578-5_1

sheep. They extracted living nuclei from some of the cells, and substituted these for the nuclei in fertilized goat ova. Dozens of such modified ova were implanted into female goats, of which seven became pregnant. Six of these goats miscarried, but one reached term. On July 30, 2003, a female kid was obtained by caesarean section from this surrogate mother. A DNA test later confirmed it to be Celia's clone (Folch et al., 2009).

Unfortunately, a lung abnormality (a common complaint in clones) caused the bucardo kid severe respiratory distress and, after a few strangled breaths, it died—just seven minutes after being born. Still, it did manage some strangled breaths. As a result, the bucardo now, arguably, has the dubious distinction of being the first animal ever to go extinct *twice*—first in 2000 when Celia was crushed by the falling branch, and then a second time, in 2003, when her clone expired.

A *de-extinction* is the reversal, or undoing, of an extinction. The concept needs little introduction thanks to Hollywood's 1993 blockbuster, *Jurassic Park*—a cautionary tale of resurrected dinosaurs running amok and killing the hubristic scientists who created them. When *Jurassic Park* first hit the movie screens de-extinction was still the stuff of science fiction, but the technology has since advanced in leaps and bounds. The 1990s witnessed the advent of mammalian cloning, and high-speed, inexpensive genome sequencing came of age in the 2000s. The 2010s are the decade of CRISPR, a game-changing gene editing technology that allows genetic engineers to cut and paste genes into chromosomes virtually at will. Biotechnology is now developing at an exponential rate, and the implications for medicine, agriculture and human society at large could hardly be more profound. So too are the implications for wildlife conservation. As the technological obstacles to de-extinction have tumbled, multiple de-extinction projects have been announced around the world, including attempts to resurrect the aurochs, the woolly mammoth, and the passenger pigeon—three species that will be the focus of Chap. 2.

The wheels of biotech spin fast. Those of philosophy turn rather more sedately. Now that 'the bucardo is out of the bag' (as it were), philosophers have an urgent game of catch-up to play. De-extinction throws up a host of controversial and important philosophical questions, the foremost of which are these:

- *The Authenticity Question.* Can de-extinction technology be used to genuinely reverse the extinction of a species, by boosting its population size from zero to a higher number?
- *The Ethical Question.* Should conservationists judge de-extinction to be ethical? (That is, should they embrace it, or not?)

This book is about the philosophy of de-extinction in general, and the answers to these two questions in particular.

1.2   THE AUTHENTICITY QUESTION

Which animal was the last bucardo to live and breathe? Was it Celia? Or was it her clone? The answer depends on what Celia's clone *was*. Was it an authentic bucardo, like Celia, or was it something else—a Frankenstein creation, an animal of a new and unnatural type?

Here is one way of understanding the relationship between Celia and her clone. Since they shared identical chromosomes they were in effect identical (monozygotic) twins, albeit twins born to different mothers far apart in time. If we think of Celia's clone in this way—as being Celia's belatedly born identical twin—then it follows that the clone was itself a bucardo, just like Celia. The last bucardo to live and breathe, therefore, was Celia's clone, not Celia.

But there is a second possibility. Celia's clone differed from Celia in at least two potentially important ways. First, the clone was not quite a perfect genetic copy of Celia. While the clone did have the same nuclear DNA (i.e., the same chromosomes) as Celia, its mitochondria came from a surrogate mother (a goat) who donated the ovum from which the clone grew, not from Celia, who contributed only the nucleus. Therefore, the clone's mitochondrial genes were goat genes, not bucardo genes. Second, Celia's clone differed from Celia with respect to her *history* and *mode of genesis*. Celia was a natural organism—the product of aeons of life, death and differential reproductive success among the bucardo of prehistoric Europe. Celia's clone, in contrast, owed her origins to a team of white-coated synthetic biologists working with machines and chemicals in a biotech lab. For either or both of these reasons—or perhaps for some other reason—we might conclude that Celia's clone was not an authentic bucardo, and that she was instead a member of some new, artificial, synthetic species, a mere 'pseudo bucardo' as it were.

If this second way of thinking about Clelia's clone is correct then the bucardo did not go extinct *twice*. It instead went extinct *just once*, when Celia met her unfortunate end. Because it wasn't truly resurrected when Celia's clone was created, it was never in a position to go extinct a second time. On this way of understanding events, Celia was the very last of her species. Her clone was the first (and perhaps the last) of a new, different, artificial breed of animal.

Here, we can distinguish *authentic de-extinctions* from *pseudo de-extinctions*.

Authentic de-extinction: increasing an extinct species' total population size from zero to some higher number, by creating new, living organisms that are authentic members of the same species.

Pseudo de-extinction: leaving an extinct species' total population size at zero, so that the species is still extinct, but creating organisms that resemble the extinct organisms closely enough to be easily mistaken for them, even though they are in fact members of a wholly new, synthetic 'race' of organisms.

In other words, an authentic de-extinction involves the true 'reversal' or 'undoing' of an extinction event, while a pseudo de-extinction does not. The products of an authentic de-extinction are the genuine articles or the real McCoys, while the products of a pseudo de-extinction are mere lookalikes, or fakes, or shams, or proxies, or facsimiles, or simulacra. A pseudo de-extinction is very much a low-grade or poor man's de-extinction. That said, a pseudo de-extinction might still be of considerable value from the conservationist perspective. For example, pseudo de-extinct organisms might be able to act as ecological proxies or surrogates for the extinct organisms they are based on, by performing important ecological functions the extinct organisms used to perform.

If authentic de-extinctions are possible then the answer to the Authenticity Question is affirmative. If pseudo de-extinctions are the best that is possible, then the answer is negative. We will give the name 'authenticism' to the doctrine that authentic de-extinctions are possible and that the answer to the Authenticity Question is affirmative. Thus an *authenticist* is someone who thinks authentic de-extinctions are possible, while an *anti-authenticist* is someone who denies this.

To see why it matters who is right—the authenticist or the anti-authenticist—consider the system used by the IUCN Red List for ranking species by conservation status. It involves assigning species to the following categories (IUCN, 2017):

Least concern
Near threatened

Vulnerable
Endangered
Critically endangered
Extinct in the wild
Extinct

If authentic de-extinctions are possible, then a new category will need to be added to the bottom of this list, namely:

Terminally extinct

If a species is extinct, yet still salvageable because it is amenable to authentic de-extinction, then it will not be terminally extinct. Rather, it will be in a kind of limbo from which biotechnology can summon it back. On the other hand, if an extinct species cannot be made authentically de-extinct—say, because any last vestiges of its DNA have long since degraded to gibberish—then it will be terminally extinct rather than merely extinct. Of course, if authentic de-extinctions are impossible—if anti-authenticists are correct and the answer to the Authenticity Question is negative—then all extinctions are terminal. There will, in this case, be no limbo betwixt life and death for a species to go into, and the old conservationist rallying cry that "Extinction is forever!" will express an iron rule with no exceptions.

The assumption that extinction is forever has long been taken as axiomatic by environmentalists. The philosopher, Holmes Rolston III, expresses the thought by equating extinction with a "superkilling":

> It kills forms (species) beyond individuals. ... It kills birth as well as death. Afterward nothing of that kind either lives or dies .... Life on Earth cannot exist without its individuals, but a lost individual is always reproducible; a lost species is never reproducible. (1991, p. 85)

In a similar vein, the naturalist, Peter Matthiessen (1959) spoke of extinction's "awesome finality".

If this widespread assumption is false—if authenticism is true and extinctions need not be forever—then conservationism's priorities will need to be rethought. Most obviously, if a species has already gone extinct but is not yet terminally extinct, we will need to ask ourselves whether we have mourned it prematurely or whether it is worth spending precious conservation resources trying to recover it. A less obvious (but probably more important) implication of authenticism's being true is how we should deal with *future extinctions*. All indications are that the surge of anthropogenic extinctions the Earth has witnessed so far is but the harbinger of a much larger Holocene mass extinction still to come. Irreversible planet-wide changes that have already been set in motion by our species—to the climate, to sea-levels, to the ocean's pH, to the distribution of species—are set to have calamitous effects over coming decades and centuries. As this mass-extinction event gathers force, conservationists will be overwhelmed. (Indeed, they are overwhelmed already.) Intensive triaging will be necessary, with desperate decisions having to be made as to which species to save and which to let slip away.

Rather than simply letting a species slip away forever, another possibility would be to let it undergo a 'managed extinction' by cryogenically preserving as many of its cell-lines as possible. By this means conservationists could, in those cases where they cannot save a species, at least prepare the ground to bring it back in future, if and when it becomes possible to do so. Conservationists living centuries hence, with technological capacities exponentially greater than ours, could be very grateful to us for our foresight if we were to put the Earth's lost biodiversity 'on ice' in this way. While it might seem strange to base what we do now on speculation about what people will be capable of in the distant future, such centuries-long timeframes—being the timeframes over which large trees grow—are very much part-and-parcel of ordinary conservationist thinking and planning.

De-extinction could be of huge value as a conservationist tool in the very long-term, by providing our species with a way of recovering some precious fraction of the biodiversity that will be lost to the coming mass-extinction event. But this assumes that authenticism is true—i.e., that de-extinction truly offers a way of *recovering lost biodiversity*, as opposed to *creating artificial biodiversity*. If authenticism is false then most of de-extinction's apparent promise as a tool for conservation is fraudulent. De-extinction has many anti-authenticist critics who have made this point in no uncertain terms. For example, Blockstein (2017) scathingly remarks that "one of the ethical violations of the proponents of 'de-extinction' is to lure and seduce the public with false promises ... of bringing extinct species back from the grave". Switek (2013) writes that "'revive and restore' projects are actually creating new species rather than truly resurrecting what was lost". And Minteer (2015) says that "de-extinction proponents too casually and uncritically equate ... engineered doppelgängers with ... vanished species".

Are these complaints fair? Is authenticism true, or false? Is the answer to the Authenticity Question affirmative, or negative? This will be the main topic of Chap. 3, where we examine arguments both for and against authenticism. There we will conclude that the pro-authenticism arguments are much more convincing than the opposing anti-authenticism arguments. In short, we will endorse authenticism.

1.3   The Ethical Question

Is de-extinction *good* or *bad*, at least as considered from the perspective of wildlife conservation? Is it something that *should* be done, or not? One view is that it is a powerful new tool in the conservationist's toolkit, that if used wisely, might enable us to undo many of the environmental harms we have already been responsible for, and mitigate many environmental harms that we are likley to cause in the future—which is therefore, perforce, good. A contrary view is that the very idea of curing environmental ills by tampering with genes and synthesising new life is symptomatic of the misguided hubris that is the root cause of these problems in the first place (Minteer, 2015). Needless to say, the Ethical and Authenticity Questions are closely connected. For instance, someone who thinks de-extinct organisms will be inauthentic is likely, for this reason, to think repopulating wilderness areas with creatures produced using de-extinction technology would be a bad idea.

The cloning of Celia stands as proof that de-extinction is possible, after some fashion at least. It also vividly illustrates reasons why de-extinction could be seen as extremely morally problematic:

- Many goats had to undergo painful and stressful veterinary procedures in order to create Celia's clone, and the clone's brief life was one of intense suffering. One objection to de-extinction—the 'animal welfare argument', as we will call it—is that it involves inflicting unnecessary suffering on animals.
- The money and resources that went into creating Celia's clone could have been much better spent preventing the bucardo going extinct in the first place. Precious and inadequate conservation funding needs to go where it will do the most good—namely to protecting extant species and their habitats. De-extinction projects could, in drawing money and talent away from such critically important conservation work, be profoundly counterproductive.
- If Celia's clone had survived, she would only have been one, lonely animal—and one animal does not make a breeding population. Even

if more such clones had been created, they would all have been genetically identical, and they would all have been females. There was, therefore, never any possibility of creating a new bucardo breeding population by cloning Celia. Moreover, even if such a breeding population *could* be created, the species would in all likelihood go promptly 're-extinct' again if the animals were released back into the wild (especially as the cause of the initial extinction, poaching, is still operative in the Pyrenean mountains). A de-extinction project will be pointless from a conservationist perspective unless it is part of a coherent larger plan to re-establish a genetically healthy breeding population of the species in a wilderness area where its future will be secure.

- In so far as the fleeting 'de-extinction' of the bucardo threatens the idea that "extinction is forever" it appears to be inimical to the larger, long-term interests of the conservation movement. By encouraging the idea that extinctions can be undone 'with the wave of a technological magic wand' it can only feed complacency about the fate of the more than 5000 species presently listed as critically endangered (IUCN, 2017). The advent of de-extinction technology will enable powerful interests to excuse extinctions with promises to undo them later, and in so doing create a dangerous moral hazard.

These are powerful arguments against de-extinction. They need to be balanced against arguments for de-extinction—including the argument that de-extinction is justified *by its being a way of restoring valuable biodiversity that has been lost*, and the argument *that we owe it as a duty to species we have exterminated to bring them back*. Chapter 4 will survey and critically evaluate all the main ethical arguments both for and against the value of de-extinction as a conservationist technique. It will conclude that arguments for de-extinction are sometimes, but not always, trumped by arguments against it, with the final ethical verdict being heavily dependent on the details of the particular de-extinction project in question. In short, it will offer *a limited ethical endorsement of de-extinction*.

1.4   Types of De-extinction

So far we have spoken only about one type of de-extinction—namely the de-extinction *of species*. Species de-extinctions will be our primary focus throughout the rest of the book, and wherever we speak of 'de-extinction' without further qualification it is species de-extinction we will be discussing. However, species de-extinction is only one of several different forms of de-extinction. This is important to recognize, partly because versions of the Authenticity and Ethical Questions arise in relation to the other types of de-extinction too, and because the answers one gives to the different versions of the questions need to be consistent.

The four main types of de-extinction are as follows (Campbell, 2016, p. 748):

1. Intraspecies gene de-extinction
2. Interspecies gene de-extinction
3. Ecosystem de-extinction
4. Species de-extinction

An intraspecies gene de-extinction is the resurrection of an *individual gene*, or a number of individual genes. A gene goes extinct when it is lost from the genepool of a species, due to there being no members of the species left alive with that gene. Resurrecting a gene is a matter of creating new living members of the species that do have that gene. This can be accomplished by inserting a copy of the gene into the genome of a fertilized ovum, and then arranging for the modified ovum to develop into a mature organism. Intraspecies gene extinction has the potential to be an immensely valuable conservation technique. For example, species that have passed through a genetic bottleneck (and thereby flirted with extinction) typically have sharply reduced genetic diversity. Such 'genetically depauperate' species suffer from diminished reproductive vigour—a phenomenon known as inbreeding depression. They also have decreased 'evolutionary potential', which is to say, a lessened ability to adapt to environmental change (Steeves, Johnson, & Hale, 2017). For these reasons, species that have been saved from extinction often remain at severe risk of extinction. Intraspecies gene de-extinction offers conservationists a chance to accomplish what would otherwise be impossible—viz., to give such species back their lost genetic diversity, thereby resolving inbreeding depression and restoring evolutionary potential. A project to apply this technique to the endangered black-footed ferret is now underway (Biello, 2016), and numerous other extant species could benefit from the same intervention.

*Intra*species gene de-extinction is to be contrasted with *inter*species gene de-extinction. As just explained, the former involves inserting a gene that has been lost from the genepool of a species back into the genepool *of the same species.* The latter is much more radical. It involves inserting an extinct gene into the genepool *of a different species.* One example, to be discussed in detail in Chap. 2, is the possibility of transferring woolly mammoth genes for cold tolerance into the genomes of Asian elephants, with the aim of creating Asian elephants that can live in the temperate zones of Eurasia. This could open up a huge new area of habitat to the Asian elephant species, enabling it to escape intense human population pressure in the tropics.

Interspecies gene de-extinction is itself a version of a more general type of conservationist intervention called *facilitated adaptation* (Thomas et al., 2013). Facilitated adaptation is human-assisted genetic adaptation. It involves genetically fortifying a species against some threat—e.g., disease, habitat loss, pollution, or climate change—by equipping the species with useful genes from another (potentially extinct) species. For example, the huge and beautiful American chestnut, once the dominant tree in the forests of the eastern United States, was almost completely wiped out early in the twentieth century by the accidental introduction of the Asian chestnut blight fungus. A transgenic strain of American chestnut has now been created that is resistant to the chestnut blight thanks to a gene from wheat that has been inserted into its genome. With the aid of this new gene the American chestnut may again dominate the forests of the eastern U.S. in the not too distant future (Jabr, 2014).

Next, *ecosystem de-extinction.* What we call an 'ecosystem de-extinction' is the creation, through genetic engineering, of an *ecological proxy* for some extinct 'keystone species'. A *keystone species* is a species that provided ecological services essential to the maintenance of some ecological system. The aim of creating such an ecological proxy would be to restore or resurrect an entire lost ecosystem and thereby benefit all the innumerable species that lived in that ecosystem. An oft-discussed example, already alluded to, would be the creation of a cold-tolerant version of the Asian elephant. The most obvious beneficiary of the elephant being re-engineered in this way would be the elephant itself, since enormous new tracts of habitat would thereby be opened up for it, in the parts of the Earth once grazed by woolly mammoths. But there would arguably be a great many other beneficiaries too. For example, due to their ability to clear trees from the landscape, woolly mammoths and other extinct grazing megafauna (such as the woolly rhinoceros) created a huge area of biologically rich grassland covering northern Siberia, Alaska and Canada's Yukon. In the mammoth's absence, these grasslands have been replaced by less productive and less biodiverse forests. Herds of cold-tolerant Asian elephants could again provide the crucial ecosystem engineering services that woolly mammoths used to provide. '[They] would be free to wander in the open spaces of Siberia, Alaska, and Northern Europe, restoring to these places all of the benefits of a large dynamic herbivore that have been missing for 8000 years' (Shapiro, 2015, p. 207). (This plan to create cold-tolerant elephants will be discussed more fully in Chap. 2.)

Finally, *species de-extinction*. Species de-extinction can be thought of as an extreme form of intraspecies gene de-extinction. The aim is to resurrect, not just one lost gene, or several lost genes, but *an entire lost genome*. It involves, not just inserting lost genes back into an extant genepool, but recreating the genepool in its entirety, from scratch.

As we have said, species de-extinctions will be our main focus in this book. Before we go any further it will be useful to introduce some distinctions between different types of species de-extinctions.

One distinction is that between what we will call *nominal de-extinctions* and *Least Concern de-extinctions*. A 'nominal de-extinction' is a species de-extinction that only barely crosses the threshold for resurrecting a species. It is a de-extinction, like the creation of Celia's clone, that yields only a single, perhaps very short-lived organism. Although a nominal de-extinction creates a new living member of an extinct species, and thereby (if perhaps only very briefly) changes the status of that species from 'extinct' to 'non-extinct', it doesn't stop the species from being *functionally extinct*, because it does not result in a viable breeding population. With respect to conservation value, nominal de-extinctions lie at the 'little value' end of a spectrum, with the other end occupied by the de-extinctions we will call 'Least Concern de-extinctions'. A Least Concern de-extinction is one that *does* result in a viable breeding population of the formerly extinct species. Not only that, but this viable breeding population is translocated into the wild, where it thrives and grows to such healthy numbers that it eventually earns the IUCN Red List status of 'Least Concern'. In other words, Least Concern de-extinctions are the most wildly successful de-extinctions imaginable.

It is obviously easy to define other types of de-extinction that fall at intermediate points in the spectrum between nominal and Least Concern de-extinctions. For example, a 'Vulnerable de-extinction' would be a de-extinction that results in the creation of a population that eventually earns the IUCN Red List status of 'Vulnerable' but that never achieves a better status than this.

A second useful distinction can be drawn between *unmanaged* and *managed* de-extinctions (Campbell, 2016, p. 748). A 'managed de-extinction' would be the de-extinction of a species that went extinct only subsequent to deliberate preparatory groundwork having been laid for its eventual de-extinction. The most helpful form of preparatory groundwork would be the cryobanking of genetically diverse live cell-lines. The existence of these preserved cells would dramatically simplify the technical hurdles that any later de-extinction project would need to overcome. An 'unmanaged de-extinction', by contrast, would be a de-extinction that takes place without the benefit of any such preparatory groundwork.[1] Since the technology of cryopreservation came of age only in the 1960s, no species that went extinct before this time will be a candidate for managed de-extinction, and only a select few species that have gone extinct since then are candidates (with Australia's extinct gastric brooding frog being one example (Archer, 2013)). However, the cryobanking of cell-lines is now commonplace thanks to the efforts of such organizations as San Diego's 'frozen zoo' (Benirschke, 1984; Ryder, McLaren, Brenner, Zhang, & Benirschke, 2000). This being so, managed de-extinctions could one day be a widely available option for species that are still extant now but that go extinct in the future.

The third division we will draw is between *precipitate* and *deferred* de-extinctions. A 'precipitate de-extinction' would be a de-extinction carried out as soon as a species goes extinct or as soon as the technology to resurrect the species become available. In contrast, a 'deferred de-extinction' would involve a deliberate delay, perhaps lasting for decades or centuries. Possible reasons for such a delay include there being other more pressing near-term conservation priorities, or anticipated improvements in de-extinction technology and its affordability. Notice that any deferred de-extinction would almost certainly be a managed de-extinction, since it would make little sense to plan to resurrect a species in the future but not to carefully cryobank all available cell-lines and DNA samples.

A fourth distinction can be made between *exacting* and *inexact* de-extinctions. A (perfectly) exacting de-extinction would involve the creation of new organisms that are completely indistinguishable in all genetic, physiological, morphological and behavioural respects from typical organisms in the pre-extinction population. An inexact de-extinction would instead involve the creation of organisms that only roughly approximate the members of the pre-extinction population in such respects. Hence exactingness comes in degrees. For example, the cloning of Celia was a relatively exacting de-extinction of bucardo, at least in so far as the clone had exactly the same nuclear DNA as Celia. That said, it was not perfectly exacting, since Celia's clone had the mitochondrial genes of a goat, not of a bucardo.

It is plausible that the distinction between exacting and inexact de-extinctions is closely related to the distinction between authentic de-extinctions and pseudo de-extinctions—since it would appear that the products of an inexact de-extinction will, in consequence of their being poor copies of the originals, be inauthentic. This is a point we will return to in Chap. 3 when we discuss the Authenticity Question.

A fifth useful distinction is between what we will call *discriminating* and *undiscriminating* de-extinctions. Every population of organisms carries a *mutation load*, this being the reduction in its reproductive fitness caused by deleterious genes in its genepool. (Such deleterious genes are constantly being added to the genepool by mutation and then gradually flushed out of it again by negative selection.) Synthetic biologists who are resurrecting an extinct species could, if they decided to, identify deleterious genes in the old genepool and deliberately exclude these genes from the genomes of the new organisms they are creating. The result would be a population of 'super organisms' that are exceptionally fit due to their having an unnaturally low mutation load. This is what we will call a 'discriminating de-extinction'. In contrast, an 'undiscriminating de-extinction' would be a de-extinction aimed at reconstructing the lost genetic diversity of the species in full, warts and all, without any attempt to select advantageous genes and omit deleterious genes. Notice that the effects of a discriminating de-extinction would gradually wear off over time (at least in the absence of further genetic interventions), because, as one generation follows another, the background mutation rate would cause the de-extinct population's mutation load to slowly creep back up to natural levels. But in the meantime, the genetic advantage enjoyed by the resurrected organisms could help them regain a foothold in an inhospitable world.

A sixth division can be drawn between *anthropogenic* and *non-anthropogenic* de-extinctions. An anthropogenic de-extinction would be the reversal of an anthropogenic (i.e., human caused) extinction. A non-anthropogenic de-extinction would instead be the reversal of a natural extinction—i.e., an extinction that humans did not instigate. In principle, it is an interesting question whether the reasons we might have for reversing anthropogenic extinctions extend to non-anthropogenic extinctions too. However, in practice this issue is largely moot because humans are implicated in virtually all extinctions that occurred recently enough for the lost genome to be recoverable.

A seventh and final division can be drawn between species de-extinctions based on the *method of resurrection* used, as follows. (See Jebari (2016) for a related discussion.)

1. Cryptobiotic de-extinctions
2. Gametic de-extinctions
3. Embryonic de-extinctions
4. Back-breeding de-extinctions
5. Clonal de-extinctions
6. Reconstructive de-extinctions
7. iPGCT de-extinctions

A *cryptobiotic de-extinction* (as we will call it) would exploit the phenomenon of cryptobiosis, which is a capacity possessed by some organisms to be revived from a completely ametabolic state. These include various plant seeds (Walters, Wheeler, & Stanwood, 2004), fly larvae (Hinton, 1960), beetle larvae (Carrasco et al., 2012; Sformo et al., 2010), caterpillars (Yi & Lee, 2016), and tardigrades (Wright, 2001) that can be frozen solid and then brought back to life. When such organisms are frozen and all their metabolic processes have come to a complete standstill there is a real sense in which they are not 'living'. If every member of such a species was to be frozen like this simultaneously, then there would be a sense in which the species would at this particular moment in time have 'no living members', making it technically 'extinct'. Resurrecting the species would then be as easy as gently thawing some of the organisms until they began metabolizing again. This would be a 'cryptobiotic de-extinction'.

A *gametic de-extinction* would be accomplished by cryobanking gametes—ova and sperm—from a species prior to its extinction and then reviving the species using *in vitro* fertilization. If the species in question fertilizes internally then the zygotes thus created would need to be implanted in the reproductive tract of a surrogate female of a closely related species in order for the embryo to develop. If it fertilizes externally then no surrogate mother would be required. (Externally fertilizing species include most amphibians, many fish, and many benthic invertebrates. Unfortunately, the ova of such species are often exceedingly difficult to cryobank, although there are exceptions (Tervit et al., 2005).) A possible variant of a gametic de-extinction would start, not with cryobanked gametes but with cryobanked somatic cells, which would be converted into pluripotent stem cells, then into primordial germ cells, and then into ova and sperm (Jebari, 2016).

An *embryonic de-extinction* would be accomplished by cryobanking the early-stage embryos of a species prior to its extinction, and then using these to revive the species (either with or without using the surrogacy services of a closely related species, depending on whether the species fertilizes internally or externally).

Backbreeding, clonal, reconstructive and *iPGCT* de-extinctions will be discussed in detail in Chap. 2, and so we will keep what we say here brief.

A 'backbreeding de-extinction' would be accomplished by selectively breeding and cross-breeding organisms belonging to extant breeds of a species in such a way as to reconstruct the genetic makeup of a lost, ancestral breed. It is reliant on the ancient breed's genes still existing in the genepool of the species' extant population.

A 'clonal de-extinction' is accomplished by inserting live cellular nuclei from cryogenically preserved cells of the extinct species into the fertilized ovum of a closely related surrogate species in place of its original nucleus, and then (if the species fertilizes internally) implanting it into the reproductive tract of a female from the surrogate species. (The full scientific term for cloning is 'somatic cell nuclear transfer' (SCNT)—but we will just call it 'cloning'.) This was the technique used on the bucardo.

All the de-extinction methods mentioned so far rely on the existence of cryopreserved cell samples. They are therefore inapplicable to any species that went extinct before the modern era of cryobanking. (The woolly mammoth is, as we will see in Chap. 2, a possible exception to this rule because of natural cryopreservation in permafrost.) In order to resurrect any extinct species for which there are no cryobanked cells it is necessary to: first, find an old bone or skin sample that contains some of its DNA; second, sequence its genome; third, sequence the genome of some closely related, extant 'guide' species; fourth, identify points of divergence between the two genomes; and fifth, genetically engineer the cells obtained from this guide species by cutting out the divergent genes and replacing them with synthesized copies of genes from the extinct species. The aim is to effectively transform the extant species' cells into living cells of the extinct species. Having been thus re-engineered, the cells can then be used to create living organisms of the extinct species by the cloning technique already described. This is what we will call a 'reconstructive de-extinction' (with the name coming from the fact that it involves reconstructing the lost genome).

Some species—including bird species—are not amenable to cloning because of peculiarities of their reproductive systems. Another method, called 'interspecies primordial germ cell transplantation', can potentially be used to resurrect such species. It involves creating chimeric organisms whose gonads produce sperm and ova containing the genes of the extinct species. These chimeras can then be bred together to produce live offspring of the extinct species. We will call de-extinctions accomplished by this method (to be described in more detail in Chap. 2) 'iPGCT de-extinctions'.

1.5   De-extinction and the Goals of Conservation

An array of different de-extinction methods have just been distinguished. Should conservationists welcome their use, or not? The answer obviously hinges on what the goals of conservationism are, and on whether de-extinction can help them be achieved. What then are the goals of the conservation movement?

One major goal is that of *biodiversity promotion*. This devolves into two sub-goals, of *protecting extant biodiversity* and of *restoring lost biodiversity*. Biodiversity is notoriously difficult to define and quantify (Maclaurin & Sterelny, 2008), but, roughly, it is the *richness* of the biological world, as measured, not only in terms of the diversity of living species, but also in terms of the diversity of genes and phenotypic traits that are found within these species, and the diversity of the ecological communities of which species are functioning parts. A region of the Earth is biodiverse just to the extent that it contains an abundance of different types of organisms integrated into an abundance of different types of self-sustaining ecological communities. Protecting extant biodiversity is thus largely a matter of stopping the biological richness of such communities being eroded or destroyed by human activities, and of thereby saving genes and species from going locally extinct. In practice, this chiefly comes down to stopping habitat fragmentation and destruction. Restoring lost biodiversity is a matter of re-establishing destroyed habitat, and can often involve reversing local extinctions of genes or species by 'assisted colonization' (i.e., by translocating organisms from one place to another). Of course, in cases where a gene, a sub-species or a species has become *globally extinct* it will not be possible to reverse its local extinction by this method. This is where de-extinction technology could potentially make a valuable contribution.

Species are not all equal where biodiversity is concerned. The species that add most to biodiversity are, first, the ones that that are most strikingly dissimilar from other species with respect to their phenotype, ecology, and phylogeny (Maclaurin & Sterelny, 2008), and second, keystone species that provide environmental services vital for sustaining an ecosystem. An especially famous and well-studied example of a keystone species is the return of grey wolves to Yellowstone National Park in the mid-1990s. The reintroduction of this apex predator, after an absence of over seven decades, precipitated a trophic cascade through the entire Yellowstone ecosystem, most especially by limiting the number and changing the behaviour of elk (Ripple & Beschta, 2012). The resultant reduced grazing pressure allowed increasing floral diversity in the transformed ecosystem that, in turn, attracted a wider range of fauna to return or to increase in number, including another keystone species, beavers. As 'ecosystem engineers', these beavers created a riparian environment that itself allowed for increased biodiversity, and which even changed the physical nature of the landscape itself by altering river flow patterns (Monbiot, 2013).

Why are conservationists *justified* in protecting and restoring biodiversity? Why is biodiversity valuable, and worth protecting and restoring? The answer is multifaceted. Biodiversity might be valuable in many different ways, including:

- *Intrinsic value*: the (putative) value that biodiversity has in and of itself, independently of its instrumental value to human beings. Whether biodiversity has intrinsic value is controversial (Bradley, 2001; Norton, 1986, pp. 135–182, 1995; Sarkar, 2005; Smith, 2016).
- *Anthropocentric value*: the instrumental value that biodiversity has to human beings, as composed of—

  - *Economic value*: biodiversity's direct monetary value (e.g., as a source of harvestable resources, or as a source of tourism dollars, or as a source of novel bioactive compounds that could be useful as pharmaceuticals).
  - *Option value*: biodiversity's potential but as yet undiscovered or unexploited economic value.
  - *Service value:* biodiversity's value as a provider of such essential environmental services as water purification, soil maintenance, carbon sequestration, pest control, nutrient recycling, and pollination (Ehrlich & Ehrlich, 1992).

– *Amenity value*: biodiversity's non-monetary value to human beings, as composed of—

Scientific value: biodiversity's value to science as a source of new knowledge and insights within such fields as evolutionary theory, genetics, biochemistry, physiology and ecology.

Cultural value: the value of biodiversity (or some component thereof, such as the existence of some particular species) as a focus of a people's spirituality, or as an element of their cultural practices, or as a symbol of their identity.

Aesthetic value: the value of biodiversity as a source of such pleasant, valuable, and educative human feelings and emotions as joy, wonder, fascination, awe, and inspiration (Norton, 1986, pp. 106–108).

Biodiversity's aesthetic value is itself divisible into various subcomponents. Firstly, and most obviously, living creatures are often *outwardly beautiful* (or, at least, wondrous and fascinating) to the human eye or ear. As Claude Lévi-Strauss put it, "Any species of bug … is an irreplaceable marvel, equal to the works of art which we religiously preserve in museums" (Ehrlich, Ehrlich, & Holdren, 1977, p. 812).

Secondly, every species has a quota of *functional beauty*, this being the beauty inherent in its unique adaptions to its environment and ecology (Cohen, 2014; Parsons, 2007). To appreciate a species' functional beauty is to be awed by how its bodily design, physiology and behaviour realize elegant and efficient solutions to the engineering problem of getting an organism's genes to be perpetuated onwards down the generations. This is the sort of beauty inherent in the efficiency with which a spider spins its web and the web catches a fly.

Thirdly, every natural species also possesses a large measure of what we will call *ecological beauty*. This is an aspect of its aesthetic and conservation value that emerges into view only when one recognizes the *evolutionary process* by which the species came into existence, and how this process connects the species to the physical environment where it evolved, and to the numerous other species—predators, prey, competitors, symbionts, and parasites—with which it has interacted in countless ways over the course of its evolutionary history. To appreciate a species' ecological beauty is to understand how its current form has been shaped by selective forces stemming from these innumerable historical interactions, and to recognize its traits as being imbued with historically derived adaptive functions. It is to understand the species as being an integral part of a fabulously rich and byzantine natural ecosystem that has arisen over unfathomable aeons of time, that is unique to one particular part of the Earth, and which, through its historical tie to that part of the Earth, augments the overall richness of Earth's biology and history. It is, so to speak, to appreciate the species as a character in an extravagantly complex and ancient ecological drama that is still playing out, and, in so doing, to appreciate how the coherency, integrity and complexity of this drama would be degraded were the species to disappear from its cast.

One way of grasping what it is for a species to *have* ecological beauty is to imagine a species that *lacks* it. Consider the 'triffid'—a species of bioengineered plant that takes over the world in John Wyndham's 1951 post-apocalyptic science fiction classic, *The Day of the Triffids*. As Wyndham's book tells it, the triffid could 'pull up its roots' and walk on three blunt 'legs'. It had a fatally poisonous, lashing sting, and an alien, vegetable intelligence. If such organisms were to be created by synthetic biologists then the biologists could presumably easily arrange for them to have outward beauty. Given their ability to move around the countryside, and to hide in hedges in order to ambush unsuspecting human passers-by, they would have a certain, grisly functional beauty too. But would the triffid have ecological beauty? No—none whatsoever. The triffids of Wyndham's book are the products of synthetic biology and human tinkering, not of aeons of natural selection. They do not stand at the end of a lineage that leads back into the deep past and that connects them to other species with which they co-evolved. Their traits are not imbued with historically derived natural functions. They are nothing but human artefacts, and as such they have no more ecological beauty than, say, a tractor. A world with triffids in it would be more biodiverse than a world without triffids, but the extra quantum of biodiversity that the triffids bring with them would be of a peculiarly synthetic and soulless kind, void of ecological beauty. Since it lacks ecological beauty, conservationists have good reason not to value biodiversity of this type.

Fourthly, a species (or a population thereof, or an ecosystem composed of many populations) might be aesthetically valuable in part because it possesses a property that we will call 'pristineness'. A population has this property just to the degree that it has remained unaffected by modern human civilization.[2] Since the impacts of our civilization now extend virtually everywhere (Halpern et al., 2008; Sanderson et al., 2002) it is plausible that few, if any, populations or ecosystems are now one hundred percent pristine: "'pristine nature,' untouched by human influences, does not exist" (Kareiva & Marvier, 2012, p. 965). Nevertheless, some populations and some ecosystems are much more pristine that others, and—at least for the time-being, before the full effects of climate change and ocean acidification are felt—there remain large tracts of the planet that are relatively pristine (Locke, 2014, pp. 156–158).

How much does pristineness matter where biodiversity's aesthetic value is concerned? Anyone who has looked through binoculars at a rare bird and noticed with a small pang of disappointment that it has a metal band on its leg will know that pristineness matters at least a little bit. The bird is somehow less aesthetically wonderful than it could have been in view of the fact that civilization has adorned its leg with a ring of aluminium. But a great deal of conservation work would appear to be premised on the assumption that pristineness doesn't matter *very much* to a species' conservation value. After all, conservation workers are part of human civilization, and so whenever and wherever they actively intervene on behalf of some endangered ecosystem or species, they cannot help but degrade the pristineness of that ecosystem or species. The metal band on the leg of a rare bird is a case in point. Birds are caught and banded in order to help conservation biologists monitor and manage their populations—with the guiding thought being that it is better for the birds to suffer this loss of pristineness than for their species to go extinct for want of our having accurate information about them. Many species would not now exist at all were it not for such interventions. The California condor, the Galapagos tortoise, the whooping crane, the grey wolf—all these species and many others have been saved from extinction by aggressive conservationist management methods. Since the existence of these species owes everything to civilization, they are not now 'pristine' at all, yet we don't appear to value these species noticeably less after having saved them than we did beforehand. It would therefore seem that the metric we use for assessing the conservation value of a population of organisms is relatively insensitive to that population's degree of pristineness. Infinitely better for a species to be 'sullied by the human hand'—the benevolent, helping hand of the conservation worker—than for it not to exist at all.

Consider the endangered Delmarva beggar's tick (*Bidens bidentoides*), which as Rolston (1985, p. 720) notes, is a plant very similar in appearance to other beggar's tick species. It is of no plausible economic value, of no service value, and of very little scientific value. It is a nuisance to people, because of the annoying way its seeds attach to hairy legs. "As far as humans are concerned, its extinction might be good riddance" (ibid.). So, why save it? One answer would be that the Delmarva beggar's tick has intrinsic value—i.e., that its existence is a good on its own, independent of its value (or disvalue) to human beings. But, as Bryan Norton has noted, such "exotic appeals to hitherto unnoticed inherent values in nature" (1995, p. 356) are "beset by enormous problems" (2003, p. 468) and "extremely difficult to explain clearly" (ibid., p. 469). Norton argues that even while environmentalists "continue discussions" about whether intrinsic value exists (1986, p. 239), they should be urgently defending biodiversity on purely anthropocentric grounds:

> To those who are uncommitted to environmentalism ... appeals to intrinsic values in nature and to the rights of nonhumans appear 'soft', 'subjective', and 'speculative'. We can accept this fact of political life without agreeing with it. Whatever the answer to the intellectual question of whether nonhuman species have intrinsic value, ... human oriented reasons carry more weight in current policy debates. (2003, p. 470)

This seems to be good advice. But if we follow it, then what are we to say about the Delmarva beggar's tick, with its paucity of economic value, service value and scientific value, and with its lack of distinguishing outward beauty? What anthropocentric grounds are there for saving it from extinction?

Consider these words, from three major proponents of the idea that natural species have intrinsic value:

> Species have value in themselves, a value neither conferred nor revocable, but springing from a species' *long evolutionary heritage* and potential .... (Soulé, 1985, p. 731, our italics)

> What humans ought to respect are dynamic life forms preserved in historical lines, vital informational processes *that persist genetically over millions of years, overleaping short-lived individuals.* (Rolston, 1985, p. 722, our italics)

> The non-humanistic value of communities and species is the simplest of all to state: they should be conserved because they exist and because this existence is itself *but the present expression of a continuing historical process of immense antiquity and majesty.* Long standing existence in Nature is deemed to carry with it the unimpeachable right to continued existence. (Ehrenfeld, 1978, pp. 207–208, our italics)

Here Soulé, Rolston and Ehrenfeld each claim that species are intrinsically valuable because of their sheer evolutionary ancientness. Maybe they are right. But even if they are wrong—even if species don't have any such thing as intrinsic value—it remains undeniable that every extant species has had it genes and traits sculpted over unthinkable stretches of time by the life-and-death struggles of untold generations of toiling organisms; by innumerable victories of survival being won against overwhelming odds down through the ages, right up until today. It also undeniable that such even the humble Delmarva beggar's tick is cast in quite a flattering light when viewed from this historical angle. We are awed by the antiquity of the Acropolis and Egypt's pyramids. But their antiquity is as nothing next to the antiquity of the DNA-based self-sustaining biochemical chain-reaction that is the Delmarva beggar's tick. To destroy the Acropolis to make way, say, for a new Trump tower would be an act of barbaric vandalism. To let the Delmarva beggar's tick go extinct, and its ancient chain reaction be quenched, for the sake of a few thousand dollars, would be an act of barbaric vandalism for the same reason.

This is all to say that if conservationists want an anthropocentric justification for saving species like the Delmarva beggar's tick, then they can get it by appealing to what we have called the species' *ecological beauty*. Ecological beauty is the anthropocentric analogue of the kind of historically-grounded intrinsic value that Soulé, Rolston and Ehrenfeld argue for. Even if the latter doesn't exist, the former does. Unfortunately, a species doesn't wear its ecological beauty on its sleeve. Perceiving and appreciating a species' ecological beauty requires understanding something of its ancient evolutionary background. This helps explain why people with little understanding of ecology or evolution might not see the value of saving the Delmarva beggar's tick or other species like it.

So much for the various elements of biodiversity's value. Now, if a species goes extinct, does de-extinction offer conservationists a way of getting back what was valuable about it?

Let's start with the extinct species' economic value, option value, service value, scientific value, outward beauty and functional beauty. There appears no reason why each of these elements of the species' value could not be fully recovered by making the species de-extinct, provided only that the de-extinction *was sufficiently exacting*. After all, a sufficiently exacting de-extinction would produce new organisms that were nigh-on indistinguishable from the organisms that died out with regards to their genetics, physiology, appearance and behaviour. Such de-extinct organisms would be every bit as outwardly beautiful as the originals. They would be just as functionally beautiful too, because they would embody all the same cunning physiological and behavioural solutions to the problem of surviving and reproducing. They would share the same economic value and option value as the originals, since they could be used to create all the same saleable products. They would be of essentially the same scientific value—since scientists could learn from them the same things they could learn from the originals. And finally, they would be of much the same service value, because (ignoring the possibility of irreversible changes having happened in the ecosystem) they could again provide whatever ecological services were formerly provided by the originals.

But what of the other elements of an extinct species' value: namely, its intrinsic value, cultural value, ecological beauty, and pristineness? Could these also be recovered by resurrecting the species?

First, pristineness. Born as they are directly of human technology, de-extinct organisms would, unfortunately, not be at all pristine. They would be 'sullied by human civilization' right down to their very DNA. But this inability of de-extinction technology to recover the pristineness of an extinct species is presumably something conservationists will be willing to tolerate, since—as explained above—pristineness is inevitably an immediate casualty every time conservationists actively intervene to save or protect some species. De-extinction is in the same boat as other conservation methods, like captive breeding or assisted migration, where its inability to preserve or restore pristineness is concerned.

Next, cultural value. This will be an important factor for only relatively few extinct species—namely, those that were of especial cultural significance to a people. One example of such a species is the extinct New Zealand wattle bird, the huia (*Heteralocha acutirostris*), which was a 'taonga' (a prized cultural treasure) of the indigenous Māori people. The huia was for Māori the most 'tapu' (sacred) of all New Zealand's flora and fauna. The bird's tail feathers, worn as symbols of rank, were reserved only for paramount chiefs. The feathers were an important item of trade, and were kept in special ornamental boxes ('waka huia'). The huia were especially precious to the people of *Ngati huia*, the Māori tribe (named after the huia) with stewardship over most of the forests where huia were found. Huia were strikingly beautiful birds, in part because of their sexual dimorphism, which was the most extreme of any bird species on Earth. Because of their beauty, they were ruthlessly hunted to extinction by European collectors, whose attitude is neatly captured in these words by the ornithologist and collector, Sir Walter Buller:

[A] pair of Huia, without uttering a sound, appeared in a tree overhead, and as they were caressing each other with their beautiful bills, a charge of No. 6 brought them both to the ground together. The incident was rather touching and I felt almost glad that the shot was not mine, although by no means loth to appropriate two fine specimens. (1888, p. 13)

The huia's extinction was a devastating cultural blow to the Ngati huia people. When the possibility of resurrecting the huia was first mooted in the 1999 the project won the Ngati huia's support (Dorey, 1999). This suggests that at least as far as the Ngati huia were concerned, de-extinction can recover the lost cultural value of an extinct species.

Next, ecological beauty. As we have seen, an organism's ecological beauty depends on its evolutionary history and on how this history connects it to its living and physical environment. But would de-extinct organisms *have* an evolutionary history? It might appear that they wouldn't. As remarked earlier, it would appear that whereas Celia owed her origins to aeons of natural selection operating on ancestral bucardo population in the Pyrenean mountains, Celia's clone instead owed her origins only to the machinations of a team of white-coated synthetic biologists working in a biotech lab. This being so, it would seem that whereas Celia had a full quotient of evolutionary history, her clone had none. The same would appear to be true for de-extinct animals generally. Rather than being the products of natural reproduction, they are made by us. They are therefore not connected to other lifeforms by the usual historical linkages of shared ancestry and heredity. They are severed from the phylogenetic tree of which all natural living things are part. For this reason they would, so it seems, be akin to the triffids mentioned above, in being *synthetic lifeforms*, void of natural history and thus void of ecological beauty.

If this is true—if the de-extinct version of a species would have none of the ecological beauty of the original species—then this would count as a very strong black mark against the conservation value of de-extinction technology. But is it true? We will return to this question in Chap. 3, when we tackle the Authenticity Question. There it will be argued that, contrary to what has just been suggested, de-extinct organisms will in fact inherit the evolutionary history of the dead organisms from which their genes have been copied. The implication is that the lost ecological beauty of extinct organisms is fully recoverable through de-extinction.

Finally, intrinsic value. As already noted, it is controversial whether species *even have* intrinsic value (Norton, 1995, 2003; Sandler, 2007). It is also controversial what would make species intrinsically valuable, supposing intrinsic value is a real quality of species. One view—the view of Soulé (1985), Rolston (1985) and Ehrenfeld (1978), mentioned above—is that a species is intrinsically valuable in virtue of its *natural-historical properties*. Since a species' natural-historical properties are also what make it ecologically beautiful, an implication is that if de-extinction can be used to recover a species' ecological beauty (as we will argue it can, in Chap. 3), then it can be used to recover the species' intrinsic value too. A rival view (Katz, 1992, 2012) is that a species is intrinsically valuable by virtue of its independence from human technology and control. If Katz's theory is correct then intrinsic value is an objectively existing, intrinsic analogue of pristineness. We have already explained why de-extinct organisms would not be pristine, and if Katz is right then they will be devoid of intrinsic value for the same reason. However, as we have already pointed out, all conservationist interventions, especially technology-heavy conservationist interventions, are subject to the same complaint.

1.6   OVERVIEW

The rest of this book is organized as follows. Chapter 2 is about some of the different extinct breeds and species that are candidates for de-extinction, the methods by which their de-extinction might be accomplished, and the philosophical questions their cases throw up. Chapters 3 and 4 are about the answers to the Authenticity Question and Ethical Question, respectively.

NOTES

1. Unmanaged de-extinctions closely correspond to what Sandler (2013) instead terms 'deep de-extinctions'.
2. We characterize 'pristineness' only in terms of *the impacts of modern human civilization*, not in terms of *all human impacts*, because if the protracted and far-reaching ecological impacts of our Palaeolithic and Neolithic forebears were factored into consideration then much of the native flora and fauna of Africa, Eurasia, Australia and the Americas would count as having little pristineness.

REFERENCES

Archer, M. A. (2013). *Second chance for Tasmanian tigers and fantastic frogs.* Presented at the TEDx DeExtinction/National Geographic, Washington, DC. Retrieved from http://longnow.org/revive/tedxdeextinction/

Benirschke, K. (1984). The frozen zoo concept. *Zoo Biology, 3*(4), 325–328. https://doi.org/10.1002/zoo.1430030405

Biello, D. (2016). De-extinction in action: Scientists consider a plan to reinject long-gone DNA into the black-footed ferret population. *Scientific American.* Retrieved from https://www.scientificamerican.com/article/de-extinction-in-action-scientists-consider-a-plan-to-reinject-long-gone-dna-into-the-black-footed-ferret-population/

Blockstein, D. E. (2017). We can't bring back the passenger pigeon: The ethics of deception around de-extinction. *Ethics, Policy & Environment, 20*(1), 33–37. https://doi.org/10.1080/21550085.2017.1291826

Bradley, B. (2001). The value of endangered species. *Journal of Value Inquiry, 35*(1), 43–58.

Buller, W. (1888). *A history of the birds of New Zealand.* London: John Van Voorst.

Campbell, D. I. (2016). A case for resurrecting lost species—Review essay of Beth Shapiro's, "How to clone a mammoth: The science of de-extinction". *Biology & Philosophy, 31*(5), 747–759. https://doi.org/10.1007/s10539-016-9534-2

Carrasco, M. A., Buechler, S. A., Arnold, R. J., Sformo, T., Barnes, B. M., & Duman, J. G. (2012). Investigating the deep supercooling ability of an Alaskan beetle, Cucujus clavipes puniceus, via high throughput proteomics. *Journal of Proteomics, 75*(4), 1220–1234. https://doi.org/10.1016/j.jprot.2011.10.034

Cohen, S. (2014). The ethics of de-extinction. *NanoEthics, 8*(2), 165–178.

Dorey, E. (1999). Huia cloned back to life? *Nature Biotechnology, 17*(8), 736.

Ehrenfeld, D. W. (1978). *The arrogance of humanism.* New York: Oxford University Press.

Ehrlich, P., Ehrlich, A., & Holdren, J. P. (1977). *Ecoscience: Population resources environment.* San Francisco: WH Freeman.

Ehrlich, P. R., & Ehrlich, A. H. (1992). The value of biodiversity. *Ambio, 21*(3), 219–226.

Folch, J., Cocero, M. J., Chesné, P., Alabart, J. L., Domínguez, V., Cognié, Y., … Vignon, X. (2009). First birth of an animal from an extinct subspecies (Capra pyrenaica pyrenaica) by cloning. *Theriogenology, 71*(6), 1026–1034. https://doi.org/10.1016/j.theriogenology.2008.11.005

Halpern, B. S., Walbridge, S., Selkoe, K. A., Kappel, C. V., Micheli, F., D'Agrosa, C., … Watson, R. (2008). A global map of human impact on marine ecosystems. *Science, 319*(5865), 948–952.

Hinton, H. E. (1960). A fly larva that tolerates dehydration and temperatures of −270[deg] to +102[deg] C. *Nature, 188*(4747), 336–337. https://doi.org/10.1038/188336a0

IUCN. (2017). The IUCN red list of threatened species. Version 2017-1. IUCN. Retrieved from www.iucnredlist.org

Jabr, F. (2014). A new generation of American chestnut trees may redefine America's forests. *Scientific American*. Retrieved from https://www.scientificamerican.com/article/chestnut-forest-a-new-generation-of-american-chestnut-trees-may-redefine-americas-forests/

Jebari, K. (2016). Should extinction be forever? *Philosophy and Technology*, *29*(3), 211–222.

Kareiva, P., & Marvier, M. (2012). What is conservation science? *BioScience*, *62*(11), 962–969. https://doi.org/10.1525/bio.2012.62.11.5

Katz, E. (1992). The call of the wild: The struggle against domination and the technological fix of nature. *Environmental Ethics*, *14*(3), 265–273.

Katz, E. (2012). Further adventures in the case against restoration. *Environmental Ethics*, *34*(1), 67–97.

Locke, H. (2014). Green postmodernism and the attempted highjacking of conservation. In G. Wuerthner, E. Crist, & T. Butler (Eds.), *Keeping the wild: Against the domestication of earth* (pp. 146–161). Washington, DC: Island Press/Center for Resource Economics. https://doi.org/10.5822/978-1-61091-559-5_13

Maclaurin, J., & Sterelny, K. (2008). *What is biodiversity?* Chicago: University of Chicago Press.

Matthiessen, P. (1959). *Wildlife in America*. New York: Viking Press.

Minteer, B. A. (2015). The perils of de-extinction. *Minding Nature*, *8*(1), 11–17.

Monbiot, G. (2013). *Feral: Searching for enchantment on the frontiers of rewilding*. London: Allen Lane.

Norton, B. G. (1986). *Why preserve natural variety?* Princeton: Princeton University Press.

Norton, B. G. (1995). Why I am not a nonanthropocentrist. *Environmental Ethics*, *17*, 341–358.

Norton, B. G. (2003). The cultural approach to conservation biology. In *Searching for sustainability: interdisciplinary essays in the philosophy of conservation biology* (pp. 467–477). Cambridge: Cambridge University Press.

Parsons, G. (2007). The aesthetic value of animals. *Environmental Ethics*, *29*(2), 151–169.

Ripple, W. J., & Beschta, R. L. (2012). Trophic cascades in yellowstone: The first 15 years after wolf reintroduction. *Biological Conservation*, *145*(1), 205–213. https://doi.org/10.1016/j.biocon.2011.11.005

Rolston, H. (1985). Duties to endangered species. *Bioscience*, *35*(11), 718–726.

———. (1991). Environmental ethics: Values in and duties to the natural world. In F. H. Bormann & S. R. Kellert (Eds.), *Ecology, economics, ethics: The broken circle* (pp. 73–96). New Haven: Yale University Press.

Ryder, O. A., McLaren, A., Brenner, S., Zhang, Y.-P., & Benirschke, K. (2000). DNA banks for endangered animal species. *Science*, *288*(5464), 275–277. https://doi.org/10.1126/science.288.5464.275

Sanderson, E. W., Jaiteh, M., Levy, M. A., Redford, K. H., Wannebo, A. V., & Woolmer, G. (2002). The human footprint and the last of the wild. *BioScience*, *52*(10), 891–904. https://doi.org/10.1641/0006-3568(2002)052[0891:THF ATL]2.0.CO;2

Sandler, R. (2007). *Character and environment. A virtue-oriented approach to environmental ethics*. New York: Columbia University Press.

Sandler, R. (2013). The ethics of reviving long extinct species. *Conservation Biology*, *28*(2), 354–360.

Sarkar, S. (2005). *Biodiversity and environmental philosophy: An introduction*. Cambridge: Cambridge University Press.

Sformo, T., Walters, K., Jeannet, K., Wowk, B., Fahy, G. M., Barnes, B. M., & Duman, J. G. (2010). Deep supercooling, vitrification and limited survival to −100°C in the Alaskan beetle Cucujus clavipes puniceus (Coleoptera: Cucujidae) larvae. *Journal of Experimental Biology*, *213*(3), 502–509. https://doi.org/10.1242/jeb.035758

Shapiro, B. (2015). *How to clone a mammoth: The science of de-extinction*. Princeton: Princeton University Press.

Smith, I. A. (2016). *The intrinsic value of endangered species*. New York: Routledge.

Soulé, M. E. (1985). What is conservation biology? *BioScience*, *35*(11), 727–734.

Steeves, T. E., Johnson, J. A., & Hale, M. L. (2017). Maximising evolutionary potential in functional proxies for extinct species: A conservation genetic perspective on de-extinction. *Functional Ecology*, *31*(5), 1032–1040. https://doi.org/10.1111/1365-2435.12843

Switek, B. (2013). Reinventing the mammoth. National Geographic. N.p. Retrieved from http://phenomena.nationalgeographic.com/2013/03/19/reinventing-the-mammoth/

Tervit, H. R., Adams, S. L., Roberts, R. D., McGowan, L. T., Pugh, P. A., Smith, J. F., & Janke, A. R. (2005). Successful cryopreservation of Pacific oyster (Crassostrea gigas) oocytes. *Cryobiology*, *51*(2), 142–151. https://doi.org/10.1016/j.cryobiol.2005.06.001

Thomas, M. A., Roemer, G. W., Donlan, C. J., Dickson, B. G., Matocq, M., & Malaney, J. (2013). Gene tweaking for conservation. *Nature*, *501*(7468), 485–486.

Walters, C., Wheeler, L., & Stanwood, P. C. (2004). Longevity of cryogenically stored seeds. *Cryobiology*, *48*(3), 229–244. https://doi.org/10.1016/j.cryobiol.2004.01.007

Wright, J. C. (2001). Cryptobiosis 300 years on from van Leuwenhoek: What have we learned about tardigrades? *Zoologischer Anzeiger—A Journal of Comparative Zoology*, *240*(3), 563–582. https://doi.org/10.1078/0044-5231-00068

Yi, S.-X., & Lee, R. E. (2016). Cold-hardening during long-term acclimation in a freeze-tolerant woolly bear caterpillar, Pyrrharctia isabella. *Journal of Experimental Biology*, *219*(1), 17–25. https://doi.org/10.1242/jeb.124875

CHAPTER 2

Three Case Studies: Aurochs, Mammoths and Passenger Pigeons

Abstract This chapter examines three prime candidates for de-extinction—namely, the *aurochs*, the *woolly mammoth*, and the *passenger pigeon*. It will be about what these animals were like, why people want to resurrect them, and the methods by which their resurrections could be accomplished.

Keywords De-extinction • Aurochs • Mammoth • Passenger pigeon

2.1   INTRODUCTION

Any discussion of the philosophy of de-extinction needs to be tethered to facts about the underlying science and technology. It also needs to be sensitive to how the technical and ethical issues raised by de-extinction vary depending on the particular characteristics of the extinct species being resurrected. The present chapter sets the scene for the rest of the book by describing three case studies, involving the *aurochs*, the *woolly mammoth* and the *passenger pigeon*. We choose these species in part because attempts to resurrect them are already underway, and in part because they vary sharply both in the techniques they involve and the philosophical questions they pose.

D.I. Campbell, P.M. Whittle, *Resurrecting Extinct Species*,
https://doi.org/10.1007/978-3-319-69578-5_2

2.2   Aurochs

Among the cave paintings of France's famous Lascaux cave are some of graceful, long-legged, bulls, with huge, sweeping horns, that prance over the cave's walls alongside horses, deer and bison. The paintings were made by Palaeolithic hunters 17,000 years ago, and the bulls were of a magnificent breed now lost to the world. They were *aurochs*, the ancient ancestors of modern cattle. (A terminological note: 'aurochs' is both the singular and the plural term for animals of the aurochs breed.) Aurochs were considerably larger than modern cattle, standing up to 1.8 m in height at the withers. Bull aurochs could weigh one and a half tonnes or more. Their horns could stretch over a metre long and, unlike those of modern cattle, curved forward like a pair of spears. They were a most formidable animal.

The aurochs ranged far beyond Europe—to Britain, which was connected to Europe by a land-bridge at the time when the Lascaux cave paintings were created, to North Africa, and across Eurasia as far as China and Korea. Archaeological and genetic analysis shows they were domesticated on two independent occasions to produce both the major taxa of modern domestic cattle—the humped zebu cattle, which were domesticated in India, and the humpless taurine cattle, domesticated in Mesopotamia. These domestications took place about 10,000 years ago. In Europe the aurochs populations began declining around 8000 years ago as their taurine descendants and the Neolithic human herders who farmed them expanded westward and northward from the Near East. Human hunting and agricultural pressures further reduced the aurochs' range over the next several thousand years, until by the thirteenth century AD they were confined to pockets of remote habitat in central and eastern Europe. The last known aurochs, a female, died in a forest in Poland in 1627. Although not recognised at the time, this was the first recorded extinction. It would be another one hundred and seventy years before Georges Cuvier would overturn the religious theory that species extinctions are impossible.

2.3   Heck Cattle: The Aurochs' Regrettable Nazi Connection

Decades before the discovery of DNA and long before anyone had even begun to dream of the techniques of modern synthetic biology, a pair of German bothers, Heinz and Lutz Heck, came up with a plan to resurrect the aurochs. The Heck brothers were motivated by a fascination with animal breeding and by a romantic interest in Teutonic myth and poetry, which celebrated the slaying of an aurochs as an act of extreme courage. Both the brothers were zoo directors (and the sons of a famous zoo director) (Driessen & Lorimer, 2016; Lorimer & Driessen, 2016). Heinz ran Munich's Hellabrunn zoo, and Lutz ran the Berlin zoo. Their plan involved *reversing* the process by which the aurochs had been domesticated. The aurochs had been selectively bred over many millennia to become more docile, to have shorter and less dangerous horns, and to be of a smaller, more manageable size and weight—with the final result being modern cattle. The Heck brothers would turn this process on its head. They would breed together cattle that had large horns, that were giants of their kind and that were aggressive, and they would keep doing this—selecting the largest, biggest-horned, most evil-tempered offspring at each stage—until they had recreated the aurochs. In Lutz Heck's words, they sought "fire, agility and bravery" (Heck, 1936, p. 255), and the aim "was to unite in a single breeding stock all those characteristics of the wild animal which are now found only separately in individual animals" (1954, p. 143). This method of recreating a lost breed of animal is now called 'back breeding'.

Under the direction of the Heck brothers, both zoos began aurochs breeding programmes in the early 1920s. The two breeding programmes proceeded independently, with each brother selecting different primitive cattle breeds as breeding stock and crossing them in different ways. Lutz used Spanish fighting bulls, for example, while Heinz did not (Faris, 2010; Keats, 2017; Landers, 2016).

In the early 1930s, at around the time of the Weimar Republic's collapse and the Nazis' ascension to power, both the brothers announced success. Neither brother had in fact created a breed of cattle as impressively proportioned as the aurochs, but both could at least claim of their cattle that they were very large, very big-horned and unruly to the point of being downright dangerous.

The brothers' lives took sharply different turns under the Nazi regime. Heinz was briefly jailed in Dachau for suspected communist sympathies and because of his earlier marriage to a Jewish woman, whereas Lutz signed up to the Nazi party and became firm friends with Herman Göring, Hitler's second-in-command. Lutz and Göring shared a keen fascination in hunting, and in animals that could be hunted. The Heck brothers' interest in restoring mythic Germanic landscapes jibed with Nazi sensibilities, as did the idea of 'racial hygiene' and the necessity of breeding back in order to recover lost genetic 'purity' (Boissoneault, 2017). The aurochs themselves played into Nazi ideals of primeval Aryan strength and prowess. Göring made Lutz the head of Germany's Nature Protection Authority in 1938 (Lorimer & Driessen, 2016). After Germany's invasion of Poland Lutz went on to oversee the looting of Warsaw zoo. He shipped all the valuable animals back to Germany and then organized a 'hunting party' of army officers to shoot the remaining Warsaw zoo animals. (With the zoo then closed, its director and his wife proceeded to use its grounds to save the lives of three hundred Jews.) Some of Lutz's cattle breed were released into Poland's enormous Białowieża Forest, as part of a planned hunting preserve for Nazi officers. This was intended by Lutz Heck and Göring to be the first step in the Germanization of the flora and fauna of the Third Reich's newly-conquered eastern (Lorimer & Driessen, 2016, p. 11).

As the war came to its calamitous close the cattle in Białowieża Forest were shot and eaten by hungry soldiers. Those in Berlin zoo perished in the bombing of the city. None of Lutz Heck's cattle breed survived. However, Heinz Heck's cattle in Munich fared better. Following the war "They languished in a few zoos and enclosures as small semidomesticated herds, where they gained a reputation for being especially hardy, able to withstand cold winters, on poor ground, with little human management" (Lorimer & Driessen, 2016, p. 6). They became known as 'Heck Cattle'.

In 1983 a group of thirty-two Heck cattle were released onto a Dutch wildlife reserve called the 'Oostvaardersplassen', where they have fended for themselves ever since. The herd now numbers in the hundreds. This early experiment in 'rewilding' has become extremely influential as a fore-runner of the many rewilding programmes that have since sprung up around Europe, especially over the last decade. The experiment was also very controversial, in part because Heck cattle remained tainted by their association, through Lutz Heck, with National Socialism.

2.4   Rewilding Europe

European land has been grazed without interruption since prehistory—originally by aurochs and then by domestic cattle descended from aurochs. However, since the 1960s this has begun to change. Areas of marginal land have been abandoned by farmers on a huge scale over a period of decades because of ongoing retrenchments in the animal husbandry indus-try (Terres, Nisini, & Anguiano, 2013). It is estimated that over thirty million hectares of European pasture and farmland, an area the size of Italy, will have been abandoned by 2030 (Rewilding Europe, n.d.). Following the withdrawal of farmers and their cattle, this land is now free from the ecological influence of large herbivores for the first time in millions of years. Biologically rich mosaics of grassland quickly turn into less biodiverse forest in the absence of grazing animals (Rewilding Europe, 2016). A solution to this problem is to rewild abandoned land by restocking it with wild grazing animals capable of looking after themselves without human management. Various organizations are now dedicated to this aim, the largest being *Rewilding Europe*, founded in 2011, which aims to rewild a million hectares by 2020. This enthusiasm for rewilding has provided a sudden, strong new impetus to the Heck brothers' old idea of resurrecting the aurochs. After all, what grazing animal would be better suited to providing the ecological services formerly provided by the aurochs than the aurochs itself?

There are two reasons not to use Heck cattle (Helmer, Kerkdijk-Otten, Widstrand, & Goderie, 2013; Lorimer & Driessen, 2016, p. 6). The first is their unfortunate association with the Nazis. The second is the fact that Heck cattle are in truth only poor approximations of the aurochs. They share rather little of the aurochs phenotype, and even less of its genotype (van Vuure, 2005). This is unsurprising, because by modern standards the Heck's attempts to back breed the aurochs were primitive in the extreme. The Hecks couldn't compare aurochs genes with the genes of different cattle breeds, since they were working long before the discovery of DNA's genetic role or the development of modern techniques for analysing DNA. They could only rely on the outward characteristics of aurochs—as divined from cave paintings and skeletal remains—in order to know what they should be aiming for in their breeding programmes. Likewise, they could only rely on the outward appearance and behaviour of extant breeds of cattle in order to select the particular stock they would use as their raw material to breed back from. In short, they could work only at the phenotypic level—the level of observable traits—not at the genetic level. Accordingly, any genetic resemblance between Heck cattle and aurochs could only be the result of good luck.

Would-be back-breeders of the aurochs are now in a far better position to breed towards an animal that is truly like an aurochs. The aurochs genome was fully sequenced in 2015 using DNA extracted from a 6750 year old British aurochs bone (Park et al., 2015). This means it is now a simple matter to compare the aurochs' genome with the genomes of extant cattle breeds to find out where they overlap and where they don't. Such analyses show, for example, that aurochs genes are especially common in some primitive British cattle breeds, which the aurochs apparently hybridized with before they died out (ibid.). Once it has been determined which aurochs genes are lurking within which cattle breeds, these particular breeds can then be carefully selectively bred and crossbred in such a way as to gradually reunite a full complement of aurochs genes back in a single population of cattle. These cattle would be close genetic approximations of aurochs and, hopefully, close physical approximations of aurochs too.

Several groups are now pursuing this strategy, the most established of which is Tauros Programme, initiated in 2008 (Helmer et al., 2013). But the strategy is not without potential limitations. An obvious possibility is that some aurochs genes might have been completely eliminated from the bovine genepool over the course of the last few thousand years. This would mean that "some auroch traits may … never be recoverable from living cattle breeds" (Shapiro, 2015, p. 105).

This would not prevent a close approximation of the aurochs being created, but it would mean that back breeding would not be a fully adequate method by itself. Genetic engineering would be required too. The procedure might be as follows. First, a near approximation of the aurochs would be created using back breeding. Then some cells would be taken from these aurochs-like cattle and genetically edited with the help of the recently developed CRISPR technology, which enables multiple DNA 'cut and paste' operations to be performed simultaneously in live cells. Using CRISPR, unwanted cattle genes would be excised from chromosomes in the cattle cells and synthesized copies of long-lost aurochs genes would be inserted in their place. Finally, the edited cells, with their new complement of genes, would be turned into full grown animals by cloning: fertilized cattle ova would have their nuclei replaced by nuclei containing the newly edited chromosomes, and the resulting ova would be implanted in cows' uteri, where they would grow into calves.

The bovine genome is large—three billion base pairs long. 'Correcting' every discrepancy between the cattle genome and the aurochs genome in these three billion base pairs would be a huge, time-consuming and enormously expensive task with existing technology. However, in principle it could be done. If it was done then the calves that emerged from the end of the gene-editing process would be genetically indistinguishable from the aurochs of former times. Their chromosomes would be composed of base pairs arranged in exactly the same order. There might still be important differences between the newly created animals and the original aurochs. For example, gene expression is partly controlled by heritable epigenetic markers attached to chromosomes, and the epigenetic markers in the newly created animals might not match those in the originals. But still, this would be beginning to get close to what in Chap. 1 we called an 'exacting de-extinction'—a de-extinction in which the newly produced organisms are indistinguishable from the originals in genetic, physiological and behavioural respects.

However, although such a relatively exacting de-extinction of the aurochs is possible in principle, it could be that it would not be appropriate in practice—not when the aim is to rewild abandoned farmland in Europe. The reason is that the aurochs were, by all accounts, formidably aggressive and dangerous animals (see, e.g., Bollongino et al., 2012). Julius Caesar's armies encountered the aurochs in the forests of Germany during his Gallic Wars. Caesar was duly impressed. In his written account of the wars he says:

> [They] are a little below the elephant in size, and of the appearance, colour, and shape of a bull. Their strength and speed are extraordinary; they spare neither man nor wild beast which they have espied. (Gallic War, 6.28)

Reintroducing animals of this description into the rewilded forests and grasslands of Europe, around peaceful villages and towns, could result in a lot of dead and maimed human picnickers and hikers. This was of no concern to the Heck brothers or to the Nazis. They admired the aurochs for its spirit, and wanted the animal recreated with its "fire" fully intact. But, in the era of institutionalized health and safety we now live in, the aurochs' aptitude for charging unexpectedly from the bushes and goring people to death is a major strike against it. The 'full aurochs' may not be wanted. What is wanted is an animal with the appearance and physiology of the ancestral aurochs, and with most of the behaviour too, but perhaps with a reduced inclination to unprovoked murder. Where aurochs are concerned there could be such a thing as *too much* authenticity.

2.5   Woolly Mammoths

The woolly mammoth needs little by way of introduction. It is among the most iconic of extinct animals. If there is a posterchild for de-extinction, then the woolly mammoth is probably it. Whatever else one might think of de-extinction, there can be no denying the thrill of the thought that woolly mammoths might one day tread the Earth again. Paleobiologist Teri Herridge, while arguing against mammoth de-extinction, admits that "for all my protests, I'd pay to see one if it was there, wouldn't you?" (2014). The paleoecologist Jacquelyn Gill, another sceptic of mammoth resurrection, makes a similar admission, writing "I would be the first person on a plane to Siberia if mammoths showed up in Pleistocene Park" (2013).

Where de-extinction is concerned the woolly mammoth's one great competitor in the public-affection stakes is the Tyrannosaurus rex of *Jurassic Park* fame. But dinosaurs died out 66 million years ago. DNA has a half-life that is sharply dependent on the ambient temperature. Even at sub-zero temperatures, when its longevity is at a maximum, it deteriorates to the point of complete information loss within 1.5 million years. And this is in hypothetical, ideal conditions (Allentoft et al., 2012). Currently the record for the oldest DNA to be successfully sequenced in the real world stands at less than 700,000 years (Hayden, 2013). With DNA's half-life being what it is, there is no realistic prospect of the T. Rex ever being resurrected (Horner, 2011).

But if the de-extinct dinosaurs of Jurassic Park will never be anything more than science fiction, 'Pleistocene Park' is a different story entirely. The woolly mammoth went extinct in Eurasia and North America only about 8000 to 10,000 years ago, with isolated populations surviving on subarctic islands until around 4000 years ago (Gill, 2013; Shapiro, 2015, p. 2). The sub-arctic, where the woolly mammoths roamed, is a vast natural refrigerator. The mammoths' bones, and the bones of other extinct animals like woolly rhinoceroses, are preserved by the millions in the permafrost. Moreover, the icy ground sometimes yields more than just DNA-laden bones. In rare cases the frozen carcases of cave lions, woolly rhinoceroses and mammoths are pulled from it in an almost completely preserved state. The meat on one such mammoth carcass was reportedly so fresh it began to bleed as it was cut from the ice (Shapiro, 2015). In cases like these there is a possibility (admittedly, a very faint possibility) of recovering not just mammoth DNA but also *viable mammoth cells* from the ice—cells capable of being used for cloning.

2.6   The Mammoth Steppe and the Earth's Climate

The 'mammoth steppe' was a huge biome that once stretched all across the northern parts of the globe, from Spain through Siberia and Alaska to Canada's Yukon. It was a vast area of cold but highly biodiverse and productive grasslands, teaming with herds of horses, bison, muskox, yaks, reindeer, woolly rhinoceroses and woolly mammoths. (How many animals were there? One estimate, based on an analysis of Siberian bones, is that there was an average of 'one mammoth, five bison, six horses, and 10 reindeer' per square kilometre (Pleistocene Park, n.d.).)

After existing for 100,000 years the mammoth steppe suddenly disappeared ten thousand years ago (Zimov, 2005). The reasons for its disappearance are controversial. One important contributor was probably climate change and attendant shifts in precipitation patterns. Evidence for this is provided by the timing of its disappearance, which coincided roughly with the onset of the present Holocene interglacial period. That said, the steppe had survived many previous cycles of glacials and interglacials unscathed. A second probable contributor was a collapse in the population size, and eventual extinction, of some of the megafauna—especially woolly mammoths—that maintained the grasslands by keeping it bare of shrubs and trees. (Anyone who has seen an elephant demolish a tree for its lunch will appreciate how such animals are able to 'engineer' an ecosystem.) A likely culprit in the extinction of the mammoths is predation by human 'mammoth hunters'. Plausibly, both factors worked together to destroy the steppe. Climate change would have put pressure on the mammoth population, making it less able to maintain its numbers in the face of human hunting. Reduced mammoth density, caused by human hunting pressure, would then have led to the rapid encroachment of forests and the vanishing of the steppe grasslands. Without the grasslands, the mammoth's fate would have been sealed.

The disappearance of the mammoth steppe, and of the mammoths that lived on it, is potentially connected to climate change in more ways than one. Because of the great productivity of the steppe's grasslands and very long period of time over which the grasslands flourished, enormously thick layers of carbon-rich material—largely composed of dead plant roots—were laid down in its soils, much of it frozen directly into the permafrost where it has remained frozen ever since. As a result, the soils of the former mammoth steppe are now one of the Earth's great carbon reservoirs. Two and a half times more carbon is sequestered in these soils than in all the world's forests put together (Zimov, 2005, p. 798). Unfortunately, anthropogenic climate change is causing the permafrost to begin to thaw, enabling the ancient plant material in it to rot. As it rots its carbon is released into the atmosphere, either as carbon dioxide if the rotting process is aerobic, or as methane if it is anaerobic. Carbon dioxide is, of course, a relatively potent greenhouse gas, and methane is an even more potent greenhouse gas still. This melting of the permafrost and the concomitant outgassing of carbon dioxide and methane are now well documented (Zimov et al., 2009).

In short, the frozen soils of the former mammoth steppe form part of what is potentially a very large and dangerous climatological positive feedback loop. Warming of the permafrost causes carbon sequestered in these soils to be released into the atmosphere as heat-trapping gases, which will in turn trigger more warming.

2.7   PLEISTOCENE PARK

The most influential proponent of resurrecting the woolly mammoth is a Russian by the name of Sergey A. Zimov. He directs Russia's Northeast Science Station and is a founder of 'Pleistocene Park', a fifty-square-mile nature reserve situated in a remote corner of north-eastern Siberia. Pleistocene Park is the site of a large-scale experiment to test a hypothesis about how the thawing of permafrost might be slowed. The hypothesis is that the permafrost can be protected from global warming by clearing the land of its forest cover and restoring the long-lost grasslands of the mammoth steppe. The hypothesis is based on two ideas. The first idea is that forests of snow-covered fir trees have a lower albedo (i.e., they are darker, and less reflective) than snow-covered grasslands. This is important, because the more sunlight that gets reflected straight back into space by white, snowy ground, the less the Earth will be warmed by it. Hence one effect of Siberia's forests being replaced with grasslands would be to reduce the amount of the sun's heat that is directly absorbed by the Siberian land mass and by the atmosphere above it.

The second idea is that large herds of grazing animals living on snowy grasslands during the winter will constantly trample the snow and break it up. This matters because snow is a powerful insulator. During the Siberian winter the air is usually far colder than the permafrost. If the ground is covered by a layer of snow then this prevents the permafrost being chilled by the frigid air above it. But if herds of animals are regularly trampling the snow and breaking the insulating layer with their hooves, then the permafrost will be chilled during the winter months, meaning it be much better able to survive the summers without thawing.

To test whether it is in fact possible to counteract the warming of the permafrost by recreating the mammoth steppe, Zimov's team have fenced Pleistocene Park and established herds of horses, reindeer, moose, yak, and muskox within it. The results so far are positive. Mosses and shrubs can't withstand the constant trampling and browsing, and so grasses take over, making the land better able to support the animals. At present the researchers are forced to simulate the effects of a mammoth by driving a tank around the park and using it to knock trees down. But clearly if the experiment of Pleistocene Park is to be replicated on a continental scale then tanks won't do! Either the woolly mammoths themselves, or animals that are similar to mammoths in their ability to survive the cold and keep trees at bay, will be needed.

2.8   How to Make a Mammoth

One way to make a mammoth would be by cloning. However this would be possible only if viable mammoth cells (or at the very least, intact mammoth chromosomes) could be recovered from some piece of frozen mammoth tissue. Unfortunately no viable mammoth cells have been found to date, although, as mentioned above, the freshness of some frozen mammoth tissue has encouraged speculation as to the possibility that they might be found (Kato et al., 2009; Wong, 2013). But even if such cells could be found, there is a second, very serious obstacle to the cloning of mammoths. In order to clone a mammoth the services of a surrogate elephant mother would be required. The Asian elephant is much more closely related to mammoths than the African elephant. Hence the Asian elephant, rather than the African elephant, would be the most suitable surrogate species. Unfortunately, however, the Asian elephant is endangered. In order to create a breeding population of mammoths, numerous surrogate elephants would be required. There would be risks to the health, life and welfare of the surrogate elephant mothers, caused by the stress of egg harvesting and embryo implantation, and the potential for harmful complications during pregnancy. Risking endangered Asian elephants in this way would clearly be ethically impermissible.

In short, the de-extinction of the mammoth via cloning is probably impossible and, even if it did turn out to be possible, unethical if Asian elephants would be impacted.

A second way to create a mammoth is by genetic engineering. The Harvard University geneticist, George Church, launched a project to do this after being persuaded of the merits of the idea by Zimov. Church's team began work in 2014, and the project has advanced quickly. As of mid-2017, the woolly mammoth's genome has been sequenced, as has the Asian elephant genome. The difference between the two genomes has been shown to consist of 1.4 million mutations, accounting for 0.04% of the genome. (Thus the Asian elephant is already 99.96% of the way to being a woolly mammoth, genetically speaking.) Church's team are now identifying these mutations' effects. They have established that of the 1.4 million mutations, only 2020 affect protein-expressing genes. Many of the other mutations probably affect gene expression. Lines of elephant cells have been cultured. The CRISPR gene editing technique has been used to snip elephant genes out of these cells and replace them with mammoth genes. 45 such cut-and-paste operations had been performed, targeting genes identified as contributing to the mammoth's cold-tolerance. These include genes for hairiness, genes for ear size (mammoths had small ears to avoid heat loss and frostbite), genes for subcutaneous fat, and genes for haemoglobin. (Mammoth haemoglobin had a mutation allowing it to convey oxygen very efficiently at low blood temperatures.)

The 45 edits that have been made to the elephant genome are only a tiny first step towards recreating the mammoth. But they are already enough to take us a long way towards creating what might be called a 'mammophant', this being a cold-tolerant version of the Asian elephant. A mammophant might be mistaken for a mammoth. The Siberian tourist board might even get away with marketing it as a mammoth. But in truth it would just be a peculiar-looking, slightly modified Asian elephant—an elephant with a shaggy coat, small ears, a thick layer of subcutaneous fat and good resistance to frostbite. Crucially, however, the fact that it wasn't really a mammoth wouldn't matter where its fitness for duty at Pleistocene Park was concerned. Provided it can survive the winter and knock down trees, it will be able to serve as an adequate ecological proxy for the woolly mammoth. For Zimov's purposes a mere ecological proxy is good enough. A perfect copy of a mammoth isn't needed.

How is the cell containing the edited 'mammophant' genome to be turned into a flesh and blood mammophant? For reasons just examined, the use of Asian elephant surrogate mothers would be very difficult to justify. Church's team are instead proposing to grow an embryo into an elephant calf '*ex vivo*' (outside a living body), in an artificial womb. This technology is still in its infancy, but developing rapidly. Researchers have succeeded in growing a human embryo for 13 days outside the womb, and stopped only because they were approaching the 14-day legal limit

(Shahbazi et al., 2016). Other researchers have reported similar success with mice and lambs (Bedzhov, Leung, Bialecka, & Zernicka-Goetz, 2014; Partridge et al., 2017; Shahbazi et al., 2016). Only time will tell whether the artificial womb approach will work. If it does work, then this would remove a major technical and ethical barrier to de-extinction. It would also provide conservationists with a very powerful new conservation technique that could potentially be applied to numerous endangered species. Suddenly it wouldn't be necessary anymore to use mother animals of a species in order to produce new baby animals of that species.

2.9   Passenger Pigeons

The passenger pigeon was an extraordinary species for reason of its sheer population size. It was perhaps the most numerous bird on Earth. It is estimated that approximately one out of every four land birds in the United States was a passenger pigeon (Blockstein, 2017, p. 34). They numbered in the billions. Their teeming flocks blocked the sun for hours as they went past. One recorded flock is thought to have been three hundred miles long and a mile wide. They nested hundreds-to-a-tree, with their nesting sites covering thousands of acres and sometimes containing well over one hundred million birds (Smithsonian Institution, 2001). In the words of the great environmentalist Aldo Leopold:

> The pigeon was a biological storm. He was the lightning that played between two opposing potentials of intolerable intensity: the fat of the land and the oxygen of the air. Yearly the feathered tempest roared up, down, and across the continent, sucking up the laden fruits of forest and prairie, burning them in a traveling blast of life. (Leopold, 1953)

The passenger pigeon was also extraordinary because of the suddenness of its disappearance. The birds were still hugely abundant in the 1850s. By the early 1870s their decline was noticeable. The last wild passenger pigeon was seen and shot in 1901, and the last captive passenger pigeon, 'Martha', died in Cincinnati zoo in 1914. The main cause of their demise is not contested. It was the industrial-scale, commercialized, rapacious slaughter of the birds in astonishing numbers for meat and feathers, a slaughter that became progressively more systematic and ruthless as railroads and the telegraph were introduced (Avery, 2014; Blockstein, 2017; Haught, 2017; O'Connor, 2015). It extended even to killing birds at their nesting sites despite the obvious unsustainability of this practice. During the last ever mass nesting of passenger pigeons, which occurred in Michigan in 1878, hunters killed 50,000 birds every day for almost five months (Smithsonian Institution, 2001).

The extirpation of the passenger pigeon, a creature so staggeringly hyper-abundant, was a shocking, watershed moment in humans' appreciation of the relative fragility of the natural world. It triggered the realization that unchecked exploitation could do very serious, lasting damage to nature, and was one of the main spurs to the birth of the American conservation movement.

2.10   The Passenger Pigeon Resurrection Plan

A project to resurrect the passenger pigeon is underway, involving some of America's top geneticists, paleogeneticists, ecologists and cell biologists (including the same George Church involved in the project to bring back the woolly mammoth). It is headed by Ben Novak. The plan is as follows (Novak, 2013):

Step 1, already completed, is to fully sequence the genome of the passenger pigeon (based on DNA obtained from the numerous museum specimens), and compare it to the genomes of other pigeons. The sequencing results show the closest living relative of the passenger pigeon to be the band-tailed pigeon.

Step 2, also completed, is to identify parts of the band-tailed pigeons' genome that need to be edited to transform it into the passenger pigeons' genome.

Step 3 is to grow band-tailed pigeon cells in a petri dish and edit the DNA within them. Band-tailed pigeon genes will be cut out of the cells' chromosomes, then artificially synthesized copies of passenger pigeon genes will be pasted in as their replacements. Once cells have been created that have a close approximation of the passenger pigeon genome, new lines of cells will be created in which different alleles of passenger pigeon genes are represented. This will be done in order to recapture the genetic diversity of the passenger pigeon species.

The three steps just described are very similar to the first few steps for creating a mammoth via genetic engineering, as described above. But at Step 4 the two de-extinction processes diverge. They diverge because, whereas mammals like the mammoth can be *cloned*, birds cannot (or at least, no one knows how to do it yet). The reason has to do with the way eggs are created within a bird's reproductive tract. First an ovum, attached to a yolk, is fertilized by a sperm. Then, as the fertilized ovum begins dividing, it and its yolk begin a tumbling journey down the circuitous course of the oviduct, with fibres of albumen (the egg's 'white') wrapping around them as they go. Eventually the yolk, now wrapped in layers of albumen, reaches the isthmus, where two layers of membrane are deposited around it. Next it passes through the shell gland (or uterus), where the eggshell and pigment are added. Then, finally, the egg is laid.

The problem for someone who wants to clone a bird is that it is exceedingly difficult to interpose in this egg-creation process. What one would *want* to do is to: (1) take a freshly fertilized ovum, complete with its attached yolk, out of a band-tailed pigeon's reproductive tract before the albumen is wrapped around it; (2) 'enucleate' the ovum (i.e., remove its original cellular nucleus); (3) put a new cellular nucleus in its place—a nucleus from one of the cells made in Step 3 that contains reconstructed passenger pigeon chromosomes; and then (4) put the now-modified ovum, complete with its new batch of chromosomes and still attached yolk, back in at the top of the oviduct, so that it can resume its journey down the oviduct and have albumen and an eggshell put around it. If this could be done then the resulting egg would contain a developing embryo with passenger pigeon genetics. But the process just described is problematic for two reasons. First, it means working with a microscopic ovum that has a large, very delicate yolk-sac attached to it. And second, the ovum and the yolk cannot be put back into the bird's reproductive tract immediately, because the ovum must be repeatedly 'activated' with electric shocks in order to make it undergo each of its first few divisions. (This is because of the major upset suffered by the ovum's delicate biochemical machinery in the course of having its nucleus removed and a new nucleus put into it. Essentially the ovum needs to be shocked back into life.) Unfortunately, this means that when the ovum and yolk are finally put back into the upper oviduct so that they can have albumen wrapped around them, the embryo will already be more advanced in its development than it should be at this part of its journey. In short, the timing of the whole egg-making process will have been badly thrown off.

No one has yet succeeded in cloning a bird for these reasons. (The mammalian reproductive tract poses no such technical challenges. In a mammal, a fertilized ovum simply lodges in the wall of the uterus, where it develops into an embryo. To clone a mammal, one: (1) removes the nucleus from a fertilized ovum and injects a new nucleus in its place; (2) 'activates' the resulting cell several times with electric shocks, until it is dividing under its own steam; and then (3) implants the collection of cells one has thereby created into the uterus. All going well, it will lodge in the uterus and develop normally.)

Fortunately, there is an alternative to cloning that can be used on birds. It involves a process called 'primordial germ cell transplantation' (PGCT). Primordial germ cells (PGCs) are cells which migrate through an embryo into its developing sex organs, where they grow into the animal's gonads (ovaries or testes). If PGCs obtained from one 'donor' embryo are injected into another 'recipient' embryo at the correct point in the recipient's development, then they will migrate to the sex organs and be incorporated into the recipient's gonads along with the recipient's own PGCs. The recipient will then grow up to be a 'chimera'—a creature whose body contains cells with two different types of genetic makeup. Most of its body will be made of one kind of cell, but its gonads will be partly composed of cells from the donor. This means that when the recipient animal grows to maturity some of its gametes (ova, or sperm) will carry the genes of the donor, rather than those of the recipient.

Incredibly, this can even work when the donor and recipient are of *different species*. For example, by injecting PGCs obtained from a chicken embryo into the bloodstream of a duck embryo, researchers have created a male duck that, despite looking like an ordinary duck, produced viable chicken sperm. These sperm, when used to inseminate a female chicken, issued in eggs from which hatched perfectly normal baby chickens (Liu et al., 2011). The same method has also been used to produce bustard offspring from chickens (Wernery et al., 2011). This form of PGCT, in which the donor and recipient are of different species, is called *interspecies* primordial germ cell transplantation' (iPGCT).

Step 4 of the plan to resurrect the passenger pigeon is to use iPGCT. The cells constructed in Step 3, complete with their newly inserted passenger pigeon genes, will be transformed into PGCs and then injected into the embryos of band-tailed pigeons. These embryos will grow up into adult birds that will look just like ordinary band-tailed pigeons, but they will in fact be chimeras—for their gonads will partly be composed of cells which contain the newly reconstructed passenger pigeon genome. When these adult chimeras are mated together, a sperm with passenger pigeon genes will fertilize an ovum with passenger pigeon genes, and so the egg that is eventually laid will house an embryo with passenger pigeon genes. When it hatches, the animal that steps out will be a passenger pigeon (or, at a minimum, a close approximation thereof), not a band-tailed pigeon. The passenger pigeon will be back!

Step 5 is to re-establish flourishing wild populations of passenger pigeons. This is, as Novak admits (Novak, 2013, p. 44), probably the most challenging step of all, in part for political and legal reasons. The passenger pigeons were, as Leopold put it, a 'biological storm'. Their giant flocks consumed everything in their paths, stripping the trees and fields bare in mile-wide swathes across rural America. Plagues of locusts pale by comparison. Farmers and orchardists of the twenty-first century are likely to be resistant to this 'biological storm' being unleashed on their crops. From an ecological perspective the passenger pigeon was a keystone species that, among other things, played a vital role in seed dispersal. From an aesthetic perspective they were one of the planet's great natural wonders. But from an agricultural perspective they are a long-eradicated and now happily forgotten pest.

A second likely source of resistance will come from concerns about authenticity. Hence, in Novak's understated words, 'The regulatory issues for releasing these birds could be quite complicated' (ibid.).

Setting these political and legal problems aside, the task of re-establishing the passenger pigeon in the wild also presents formidable animal husbandry challenges. Somehow the first generation of passenger pigeons need to acquire competency in the fine art of being a passenger pigeon without having any passenger pigeon role-models to learn from. Novak's ingenuous solution to this problem involves deploying a trained army of band-tailed pigeons and homing pigeon stooges that have been cosmetically dyed to look like passenger pigeons. Passenger pigeon chicks will be raised in large numbers by surrogate band-tailed pigeons that have been dyed in passenger pigeon colours. This will prevent the chicks from imprinting on the wrong species. The surrogate parents will have been 'bred and trained several generations in advance to eat the same diet as the passenger pigeon' (ibid.). They will be kept in large outdoor aviaries that simulate a forest environment. Because passenger pigeons were a nomadic species they must be taught to migrate from place to place. This will be accomplished by using dyed homing pigeons to lead them from one aviary to another. At some point the passenger pigeons will take to the forests and start nesting, or nest in the aviaries. The second generation of passenger pigeons won't need surrogate birds to look after them or teach them the lie of the land, and so the de-extinction project will then be complete.

REFERENCES

Allentoft, M. E., Collins, M., Harker, D., Haile, J., Oskam, C. L., Hale, M. L., … Bunce, M. (2012). The half-life of DNA in bone: Measuring decay kinetics in 158 dated fossils. *Proceedings of the Royal Society of London B: Biological Sciences.* https://doi.org/10.1098/rspb.2012.1745

Avery, M. (2014). *A message from Martha: The extinction of the passenger pigeon and its relevance today.* New York: Bloomsbury.

Bedzhov, I., Leung, C. Y., Bialecka, M., & Zernicka-Goetz, M. (2014). In vitro culture of mouse blastocysts beyond the implantation stages. *Nature Protocols, 9*(12), 2732–2739.

Blockstein, D. E. (2017). We can't bring back the passenger pigeon: The ethics of deception around de-extinction. *Ethics, Policy & Environment, 20*(1), 33–37. https://doi.org/10.1080/21550085.2017.1291826

Boissoneault, L. (2017, March 31). When the Nazis tried to bring animals back from extinction. *Smithsonian Magazine.* Retrieved from http://www.smithsonianmag.com/history/when-nazis-tried-bring-animals-back-extinction-180962739/

Bollongino, R., Burger, J., Powell, A., Mashkour, M., Vigne, J.-D., & Thomas, M. G. (2012). Modern taurine cattle descended from small number of near-eastern founders. *Molecular Biology and Evolution, 29*(9), 2101–2104. https://doi.org/10.1093/molbev/mss092

Driessen, C., & Lorimer, J. (2016). Back-breeding the aurochs: The Heck brothers, National Socialism and imagined geographies for nonhuman Lebensraum. In P. Giaccaria & C. Minca (Eds.), *Hitler's geographies* (pp. 138–157). Chicago: University of Chicago Press.

Faris, S. (2010, February 12). Breeding ancient cattle back from extinction. *Time.* Retrieved from http://content.time.com/time/health/article/0,8599,1961918,00.html

Gill, J. (2013, March 18). Cloning woolly mammoths: It's the ecology, stupid. *Scientific American.* Retrieved from https://blogs.scientificamerican.com/guest-blog/cloning-woolly-mammoths-its-the-ecology-stupid/

Haught, P. (2017). Integral value and the virtue of hospitality: A response to Kasperbauer. *Ethics, Policy & Environment, 20*(1), 29–32. https://doi.org/10.1080/21550085.2017.1291830

Hayden, E. C. (2013). First horses arose 4 million years ago. *Nature News.* https://doi.org/10.1038/nature.2013.13261

Heck, L. (1936). Ueber die Rueckzuechtung des Urs [On the backbreeding of the Aurochs]. *Das Tier Und Wir, 7.*

———. (1954). *Animals, my adventure.* London: Methuen.

Helmer, W., Kerkdijk-Otten, H., Widstrand, S., & Goderie, R. (2013). *The aurochs, born to be wild.* Zutphen: Roodbont Publishers.

Herridge, T. (2014, November 18). Mammoths are a huge part of my life. But cloning them is wrong. *The Guardian.* Retrieved from https://www.theguardian.com/commentisfree/2014/nov/18/mammoth-cloning-wrong-save-endangered-elephants

Horner, J. (2011). *Building a dinosaur from a chicken.* Presented at the TED2011, Long Beach, CA. Retrieved from https://www.ted.com/talks/jack_horner_building_a_dinosaur_from_a_chicken

Kato, H., Anzai, M., Mitani, T., Morita, M., Nishiyama, Y., Nakao, A., ... Iritani, A. (2009). Recovery of cell nuclei from 15,000 years old mammoth tissues and its injection into mouse enucleated matured oocytes. *Proceedings of the Japan Academy. Series B, Physical and Biological Sciences, 85*(7), 240–247. https://doi.org/10.2183/pjab.85.240

Keats, J. (2017, March). Return of the Aurochs. *Discover.*

Landers, J. (2016, April 4). Scientists seek to resurrect the aurochs, the extinct beast that inspired cave paintings. *Washington Post.*

Leopold, A. (1953). On a monument to the pigeon. In *A sand county almanac.* New York: Oxford University Press.

Liu, C., Khazanehdari, K. A., Baskar, V., Saleem, S., Kinne, J., Wernery, U., & Chang, I.-K. (2011). Production of chicken progeny (gallus gallus domesticus) from interspecies germline chimeric duck (Anas domesticus) by primordial germ cell transfer. *Biology of Reproduction, 86*(4), Art. 101, 1–8. https://doi.org/10.1095/biolreprod.111.094409

Lorimer, J., & Driessen, C. (2016). From "Nazi cows" to cosmopolitan "ecological engineers": Specifying rewilding through a history of Heck cattle. *Annals of the American Association of Geographers, 106*(3), 631–652. https://doi.org/10.1080/00045608.2015.1115332

Novak, B. (2013, September/October). The great comeback: Bringing a species back from extinction. *The Futurist,* 5 pages.

O'Connor, M. R. (2015). *Resurrection science: Conservation, de-extinction and the precarious future of wild things.* New York: St. Martin's Press.

Park, S. D. E., Magee, D. A., McGettigan, P. A., Teasdale, M. D., Edwards, C. J., Lohan, A. J., ... MacHugh, D. E. (2015). Genome sequencing of the extinct Eurasian wild aurochs, bos primigenius, illuminates the phylogeography and evolution of cattle. *Genome Biology, 16*(1), 234. https://doi.org/10.1186/s13059-015-0790-2

Partridge, E. A., Davey, M. G., Hornick, M. A., McGovern, P. E., Mejaddam, A. Y., Vrecenak, J. D., ... Flake, A. W. (2017, April 25). An extra-uterine system to physiologically support the extreme premature lamb. *Nature Communications, 8,* 15112. https://doi.org/10.1038/ncomms15112

Pleistocene Park. (n.d.). Scientific background. Pleistocene Park. Retrieved from http://www.pleistocenepark.ru/en/background/

Rewilding Europe. (2016). 2015 Annual review. Rewilding Europe. Retrieved from https://www.rewildingeurope.com/wp-content/uploads/publications/rewilding-europe-annual-review-2015/offline/download.pdf

Rewilding Europe. (n.d.). Urbanisation and land abandonment. Retrieved from https://www.rewildingeurope.com/about/background-and-goals/urbanisation-and-land-abandonment/

Shahbazi, M. N., Jedrusik, A., Vuoristo, S., Recher, G., Hupalowska, A., Bolton, V., ... Zernicka-Goetz, M. (2016). Self-organization of the human embryo in the absence of maternal tissues. *Nature Cell Biology, 18*(6), 700–708.

Shapiro, B. (2015). *How to clone a mammoth: The science of de-extinction.* Princeton: Princeton University Press.

Smithsonian Institution. (2001, March). The passenger pigeon. In *Encyclopedia Smithsonian.* Smithsonian Institution. Retrieved from https://www.si.edu/Encyclopedia_SI/nmnh/passpig.htm

Terres, J. M., Nisini, L., & Anguiano, E. (2013). Assessing the risk of farmland abandonment in the EU—Final report. EUR 25783 EN, Joint Research Centre of the European Commission.

van Vuure, C. (2005). *Retracing the aurochs: History, morphology and ecology of an extinct wild ox.* Sofia: Pensoft Publishers.

Wernery, U., Liu, C., Baskar, V., Guerineche, Z., Khazanehdari, K. A., Saleem, S., ... Chang, I.-K. (2011). Primordial germ cell-mediated chimera technology produces viable pure-line houbara bustard offspring: Potential for repopulating an endangered species. *PLoS One, 5*(12), 1–8. https://doi.org/10.1371/journal.pone.0015824

Wong, K. (2013). Fact-checking a frozen mammoth. *Scientific American.* Retrieved from https://www.scientificamerican.com/article/fact-checking-a-frozen-mammoth/

Zimov, N. S., Zimov, S. A., Zimova, A. E., Zimova, G. M., Chuprynin, V. I., & Chapin, F. S. (2009). Carbon storage in permafrost and soils of the mammoth tundra-steppe biome: Role in the global carbon budget. *Geophysical Research Letters, 36*(2), n/a–n/a. https://doi.org/10.1029/2008GL036332

Zimov, S. A. (2005). Pleistocene Park: Return of the mammoth's ecosystem. *Science, 308*(5723), 796–798. https://doi.org/10.1126/science.1113442

CHAPTER 3

Real or Fake? The Authenticity Question

Abstract Is the resurrection of an extinct species genuinely possible, or not? Will organisms produced by de-extinction technology be authentic new members of the species that died out, or just convincing fakes? We seek to answer these questions in this chapter. Critics of de-extinction have offered many reasons for thinking that the products of de-extinction will be inauthentic. The bulk of the chapter is taken up with surveying their arguments. We attempt to show that none are convincing. We end the chapter by offering and defending two arguments *in favour* of the view that authentic de-extinctions are possible.

Keywords De-extinction • Authenticity • Conservation value

3.1 INTRODUCTION

If it looks like a duck, swims like a duck and quacks like a duck, then it's a duck—or at least, so the saying goes. But if it looks like a mammoth, behaves like a mammoth, and has a close approximation of a mammoth's genome, then is it a mammoth? Might it not instead just be a mere fake or sham or faux or neo- mammoth, a creature that could easily be mistaken for a mammoth by someone unfamiliar with the creature's mode of creation, but that actually belongs to a new, artificial, synthetic species? This question—the 'Authenticity Question'—is the topic of the present chapter.

© The Author(s) 2017
D.I. Campbell, P.M. Whittle, *Resurrecting Extinct Species*,
https://doi.org/10.1007/978-3-319-69578-5_3

The Authenticity Question. Can de-extinction technology be used to genuinely reverse the extinction of a species, by boosting its population size from zero to a higher number?

Recall from Chap. 1 that this question is answered in the affirmative by *authenticists*, and in the negative by *anti-authenticists*. Authenticists believe that 'authentic de-extinctions' will soon become possible if they are not possible already. They believe organisms created using de-extinction technologies can genuinely belong to a species that was previously extinct. Anti-authenticists beg to differ. They believe it will be impossible for synthetic biologists to ever achieve anything better than a 'pseudo de-extinction', in which the originals are replaced by mere synthetic proxies, replicas, or lookalikes. So, for example, authenticists will be open to the claim that a de-extinct woolly mammoth is an authentic member of the ancient *Mammuthus primigenius* species. (Admittedly, they will probably want to know the details before fully buying into the claim. How was the de-extinction accomplished? How physically similar is the newly created animal to the mammoths of yore?) Anti-authenticists will, in contrast, dismiss this claim out of hand. They will insist the newly created animal can be no better than a pseudo-mammoth.

Who is right, the authenticists or the anti-authenticists? Is the answer to the Authenticity Question 'yes' or 'no'? Are attempts to truly 'bring back' extinct species futile, or not? To anticipate, our answers to these questions will be that the authenticists are right, and anti-authenticists wrong; that the answer to the Authenticity Question is 'yes', not 'no'; that attempts to bring back extinct species are not futile.

Chapter Overview. §3.2 begins by clarifying the point of disagreement between authenticists and anti-authenticists. §3.3–3.9 then survey the main anti-authenticist arguments, and explain why each is weak and unpersuasive. Finally, §3.10 and §3.11 present two pro-authenticist arguments, both of which we believe to be forceful.

3.2   Authenticism and Conservation Value

Before we turn to the task of evaluating arguments for and against authenticism, we will start by clarifying exactly what authenticism *says*.

On a first pass, authenticism is simply the view that the Authenticity Question is to be answered affirmatively. In other words, authenticists assert that A1 is true, while anti-authenticists assert it is false:

A1. There is some species, *S*, such that if *S* has gone extinct, or if *S* were to go extinct, then de-extinction technology could be used to genuinely reverse *S*'s extinction, by boosting its population size from zero to a higher number.

But as it stands A1 is ambiguous and open to different interpretations. It speaks of the *population size* of a species at a time, but in order to measure a species' population size we first need a *species membership criterion*—a criterion laying down the conditions it is necessary and sufficient for an organism to satisfy in order for it to belong to a given species. If there were some single, uncontested species membership criterion that we could plug into A1 at this juncture then all would be well. Unfortunately, this is very far from being the case. Within the philosophy of biology, the so-called 'species problem' is the notoriously difficult problem of defining what a 'species' actually is. Biologists and philosophers have wrestled with this problem for decades, with little success. Indeed, at latest count no less than 26 different versions of the species concept have been distinguished (Hausdorf, 2011; Wilkins, 2009), each with a different definition, each designed to serve a somewhat different operational purpose, and each associated with its own species membership criterion. There is an ongoing, bitter dispute among advocates of these different species concepts as to the respective pros and cons of each (Wheeler & Meier, 2000).

Because of the species problem, A1 needs clarifying. As a step towards clarifying it, consider the following story.

Once upon a time there was a beautiful species, *S*. It went into decline, and finally the day came when it seemed to have gone completely extinct. Not a single member of *S* could be found anywhere. So, a team of synthetic biologists tried to make *S* de-extinct. They created a flourishing new population of de-extinct organisms, called '*D*'. The members of *D* were genetically, physiologically and behaviourally very similar to the bygone members of *S*.

Had the synthetic biologists thereby *succeeded* in making S de-extinct? Here opinions were sharply divided. Some people—authenticists—were satisfied that the organisms in *D* were living, breathing members of the *S* species, while other people—anti-authenticists—were convinced that the organisms in *D* were nothing better than synthetic fakes or proxies.

But then, to everyone's astonishment, it turned out that *S* had never actually gone extinct in the first place. A hidden valley was found a thousand miles away inhabited by a flourishing remnant population of *S* organisms—'*R*', as we will call it.

At this point the authenticists believed that *two* flourishing *S* populations existed, the de-extinct population, *D*, and the newly discovered remnant population, *R*. The anti-authenticists, on the other hand, believed *R* to be the *only* authentic *S* population, for they remained steadfast in their conviction that the *D* organisms were inauthentic. (It would, after all, have been very strange for the anti-authenticists, who previously thought that the *D* organisms were inauthentic, to switch to thinking they were authentic, based not on any new information about these organisms themselves, but only on the discovery of *R* in a secret valley a thousand miles away.)

But then disaster struck! A volcano erupted near the secret valley. *R* was threatened with imminent annihilation. Could some of the *R* organisms be saved? Maybe, but it would be immensely dangerous work for the rescuers. Was the risk to human life and limb worth it? The authenticists said 'no', for in their view the continued survival of the *S* species was not dependent on the continued survival of *R*—not while *D* still flourished. The anti-authenticists vehemently disagreed, for they believed that *S* would go extinct (for real, this time) if *R* was destroyed.

The moral of this story is simply that the point of disagreement between authenticists and anti-authenticists can be recast in terms of *the conservation value of de-extinct populations*. Authenticists think a de-extinct population, *D*, could be an *acceptable substitute* for a putatively existing remnant population, *R*, where its conservation value is concerned. In other words, they think that *D* could be of conservation value in approximately all the ways that *R* (if it existed) would be of conservation value, and that the creation of *D* would therefore be a roughly adequate compensation for the loss of *R* (from a conservationist perspective, at least). Anti-authenticists are of the opposite opinion. They think *D* could have nowhere near the same conservation value as *R*, and that *D*'s continued existence would be an altogether inadequate compensation for *R*'s loss. Authenticists would be willing to trade *R* for *D*. Anti-authenticists wouldn't.

With this new way of drawing the battle-lines between authenticists and anti-authenticists in hand, we can reformulate authenticism as follows:

A2.   There is some species, *S*, such that if *S* was believed to have gone extinct, then de-extinction technology could be used to create a population of organisms that would be an adequate substitute, conservation-wise, for any remnant population of *S* organisms that might still be flourishing secretly in the wild.

Suppose that some species, S, (say, the passenger pigeon) has gone extinct, and that synthetic biologists try to resurrect it by creating a new population, D. Does the fact of D's creation mean that A2 is true? One is to answer this question by engaging in a thought experiment like the one above. One is to imagine that S has not really gone extinct after all, and that a flourishing wild, remnant population of organisms, R, descended from the old members of S, still exists 'out there' somewhere, in an unexplored corner of the world. (E.g., one is to imagine that there is a colony of passenger pigeons lurking unnoticed somewhere in the forests of Michigan.) One is then to ask oneself whether D would be approximately equivalent to R in all the respects that contribute importantly to R's conservation value. If the answer to this question is affirmative—if D is approximately 'as good' as R, and hence an adequate substitute for R, conservation-wise—then A2 is true and so authenticism, so conceived, will stand vindicated. Alternatively, if R will have some property D lacks that is a major contributor to its conservation value, then the creation of D will not amount to an existence proof of A2. (A2 might still be true in this latter case, but authenticists would need to look elsewhere, to some other de-extinction attempt, possibly involving a different species or a different de-extinction method, in order to show it was true.)

So far, two things have been established: first, that A1 is ambiguous because of the species problem; and second, that A2 provides a possible alternative to A1. Now we reach the all-important question: does A2 *improve on* A1, *by avoiding the species problem*? Later in this chapter we will argue that A2 *does* avoid the species problem, but for now we only make a weaker point: namely, that it is at least possible that it *might* avoid the species problem. One possibility, of course, is that the facts about whether D is an adequate substitute, conservation-wise, for R, will turn out to depend on prior facts *about whether or not they are two populations of one species*. The species problem would then immediately rear its ugly head again and so we would be back at square one. In this case A2 would be no improvement over A1. But there is also a second possibility: viz., that each of the various qualities of D and R that contribute importantly to their respective levels of conservation value can be assessed *without prejudging whether D and R are conspecific populations*. If this second possibility obtains—and we will argue at the end of this chapter that it does—then the species problem will be avoided by moving from A1 to A2. In this case A2 will amount to a *disambiguated* version of A1, in which the operative species membership criterion is cached out in terms of the conservation value of de-extinct organisms.

A2 is the formulation of authenticism we will work with henceforth. But before we move on, it is worth briefly noting that it sets the bar for authentic de-extinctions fairly high. It requires an authentically de-extinct population to have approximately the same conservation value as a *flourishing, wild remnant population*. The bar could quite reasonably be set lower—for example, as per A3:

A3. There is some species, *S*, such that if *S* was believed to have gone extinct, then de-extinction technology could be used to create a population of organisms that would be an adequate substitute, conservation-wise, for *a badly inbred remnant population of S organisms that are being secretly kept in captivity*.

Obviously a badly inbred remnant population of *S* organisms living in captivity would have a lot less conservation value than a flourishing wild population of *S* organisms—but still, they would be authentic *S* organisms for all that. (They would be *poor examples* of the *S* species, it is true. But they wouldn't even be poor examples of the *S* species if they didn't genuinely belong to the *S* species.) This being so, it could easily be argued that a de-extinct population need only reach this lower bar in order to count as being authentic. But we will ignore this point in what follows, since in reality the kind of authentic de-extinctions conservationists are interested in are the kind that meet the higher standard of A2. When we argue, later in this chapter, that authenticism is true, we will be arguing that this higher standard can be met.

3.3   The Genesis Argument Against Authenticism[1]

If a de-extinct animal looks like a mammoth, behaves like a mammoth, and has a close approximation of a mammoth's genome, then *what good grounds could there be for denying it is truly a mammoth?* The anti-authenticist owes the authenticist an answer to this question. One obvious possibility, to be examined over the next few sections, is that has *the wrong history*, or *the wrong mode of genesis*, or *the wrong phylogenetic relation to prehistoric mammoths*, to be an authentic mammoth.

Here are a pair of examples which illustrate how a thing's authenticity (or lack thereof) can depend on its historical properties:

Example 1. Harry accidentally sets fire to a famous Picasso painting, and it is completely destroyed. Harry is very sorry about the accident and wants to make amends, so he replaces the burnt Picasso with a copy. He goes to considerable pains to ensure that the copy is physically indistinguishable from the original. Has Harry thereby undone the harm he did? No! Although the replica could easily be mistaken for the original by someone who didn't know better, it will in truth hold barely any of the original's value. Of copies there can be many, but the original was unique. It was painted by Picasso's own hand. It was hung in famous galleries where it cast its spell over thousands of people and influenced generations of important artists. Its value inhered in these facts about its social history. The replica, in contrast, has no such history. Despite its close physical resemblance to the original, its provenance and back-story are radically unlike the original's. For this reason, it is, sadly, no better than a 'cheap knockoff'.

Example 2. There exists some beautiful natural landscape, of ancient canyons, weatherworn pillars and arches of rock, of fossils, caves, multi-coloured rock strata, and of wind-sculpted trees and bushes. This landscape is destroyed by explosives and earthmoving machinery in order to strip-mine a valuable seam of mineral ore buried beneath it. But after the ore has been removed the landscape is painstakingly restored. Coloured layers of rock, complete with inlaid facsimiles of fossils, are pasted over the land using a new rock-extrusion technology. This new rock is carved into canyons, pillars, arches and caves by enormous terraforming rock-grinding machines. Soil is trucked in and a team of landscape gardeners plant new trees and bushes that have been especially grown and sculpted to give them the same windswept look as the originals. The end result is a reconstructed landscape indistinguishable from the original, ancient landscape.

Has all the harm the strip-miners did thereby been successfully reversed and undone? No. The original landscape's coloured rock strata were created millions of years ago as sediments at the bottom of some Jurassic sea. Its canyons had been slowly carved through the solid rock by tumbling streams over unfathomable stretches of geologic time. Nothing similar will be true of the restored landscape. Its rock strata and canyons are of recent origins, laid and created by machines, in accordance with the whims and intentions of human designers. Whereas the original landscape stood as magnificent testament to the immense age of the Earth and the wonderful creative power of blind natural forces, the new landscape will instead stand as testament to human technological prowess, not to mention our greed and hubris. Hence in creating the latter we will by no means have regained what we lost when we destroyed the former. The value of the original landscape largely inhered in its history and mode of genesis. Since the restored landscape has a radically different history and mode of genesis, it will not be as valuable as the original, or, at least, not valuable in anything like the same ways. It will not be 'the same thing'. It will not be authentic.

The *locus classicus* of the idea that we cannot fully repair the environmental harm we do when we destroy a natural landscape by restoring it back to its original state is Robert Elliot's (1982) *Faking Nature*. Elliot calls the idea that the value of a destroyed landscape can be fully compensated by restoring it, the 'restoration thesis'. He rejects the restoration thesis for the reasons just alluded to. Viz., he holds that even a *perfectly* restored landscape, that was exactly like the original, destroyed landscape down to the minutest detail, would still have a radically different history than the original, and he holds that we commonly value landscapes precisely because of their *natural* origins—because 'they are unsullied by the human hand'.

Now, de-extinction concerns the artificial restoration of *destroyed natural species*, not the artificial restoration of *destroyed natural landscapes*, or *the replacement of a destroyed artwork with a replica*. However, resurrecting an extinct species would appear to be the zoological or botanical equivalent of restoring a destroyed landscape, or of substituting a replica in place of a destroyed Picasso (Cohen, 2014; Gunn, 1991; Jebari, 2016). An argument from analogy therefore suggests that lessons learned from the one type of case will carry over to the other. To develop this analogy, let's define the following thesis:

> *Resurrection thesis.* The conservation value lost when a species goes extinct can be completely or almost completely recouped by using de-extinction technology to create a new population of living organisms that closely resemble the originals in genetic, physiological and behavioural respects.

The resurrection thesis is a direct analogue of Elliot's restoration thesis. Whereas the restoration thesis is about the value of *restored landscapes*, the resurrection thesis is instead about the value of *resurrected species*. But if the analogy 'goes through'—i.e., if extinct species are truly analogous in all the relevant respects to destroyed natural landscapes—then Elliot's history-based argument against the restoration thesis will refute the resurrection thesis, too.

Does the analogy go through? On first sight, it would appear so, there being strong reasons to think that natural species are similar to natural landscapes and artworks in respect of having a value that is a product of their history. To see this, recall what was said in Chap. 1 about the different elements of a species' conservation value. Of the various elements mentioned, one of the most important was a species' *ecological beauty*— this being the aspect of its aesthetic value that inheres in our appreciating how its traits have been sculpted by millions of years of interactions with other species and with the physical environment. Would de-extinct organisms possess the ecological beauty of the originals? Seemingly not. They will, so it seems, have originated from the drawing boards, DNA synthesizing machines, and intentions (good, or otherwise) of a team of synthetic biologists. They will owe their genesis, not to the impersonal forces of evolution operating over enormous stretches of time in the natural world, *but to us*. They will be quintessentially artificial, having been created by the human hand right down to their very genetic building-blocks. Or, at least, so it would certainly appear.

In short, just as a restored version of a destroyed natural landscape would lack the *natural history* of the original landscape, and hence lack much of the aesthetic value of the original landscape, so too a de-extinct version of, say, the woolly mammoth would (apparently) lack the *evolutionary history* of the original mammoths, and therefore lack their ecological beauty.

Writing of a prospective future in which genetically engineered rabbits are widespread, Bill McKibben asks, "Why would we have any more reverence or affection for such a rabbit than we would for a Coke bottle?" (1989, p. 211). Paraphrasing McKibben, we might ask why we should have reverence or affection for a de-extinct 'mammoth' that is shorn of the evolutionary origins, and thus the ecological beauty, of ancient mammoths. Such a mammoth would—so it seems—be a mere manmade artefact, on a par with a Coke bottle. Hence (it seems) the resurrection thesis is false. Contrary to what the resurrection thesis implies, the loss of conservation value associated with a species' extinction will *not* be mostly or fully recouped by resurrecting it, the ecological beauty of the extinct species having been lost for good.

We call this the 'genesis argument' against the resurrection thesis (Campbell, 2017). The name comes from the idea that resurrected organisms lack the value of the originals because they have *an artificial rather than natural mode of genesis*. Here is a formalized version:

G1.   Organisms created using de-extinction technology will lack the evolutionary history possessed by the original members of the extinct species.

G2.   The ecological beauty of an organism depends on that organism's evolutionary history

G3.   The ecological beauty of an organism is a major ingredient of its conservation value.

G4.   If organisms created using de-extinction technology will lack a major ingredient of an extinct species' conservation value, then the resurrection thesis is false.

G5.   Therefore, the resurrection thesis is false.

Here are two examples of this argument:

> I think … de-extinction proponents too casually and uncritically equate the engineered doppelgängers with the vanished species. Their remarks certainly seem to suggest they think that the introduction of the former somehow recovers all of the values lost with the disappearance of the latter. You don't have to be an essentialist about the "natural," however, or cling to outmoded notions of species purity to recognize that there are, as we might say, morally significant differences between the extinct species and the synthesized versions. One key distinction hinges on the co-evolutionary natural history of the lost forms. Although the engineered reproductions may hold other values for conservationists, unlike their progenitors they will not have evolved in relationship to other species within a natural habitat over millennia. And that unique co-evolutionary and ecological narrative is, I believe, an important part of how and why we value wild species. It's a character that simply can't be recreated in a modern genomics lab. (Minteer, 2015, pp. 14–15)

> Many of the types of value that species possess are dependent upon their evolutionary and ecological relationships. This is the case on accounts that emphasize their natural-historical properties … Deep de-extinction, if successful, does not restore these relationships. (Sandler, 2013, p. 356)

Before we end this section, there is a technical point that needs flagging. As we have just described it, the genesis argument has *the resurrection thesis* as its target, not *authenticism*. However, the resurrection thesis and authenticism are very closely related claims, and it is but a short step from the conclusion that the resurrection thesis is false to the further conclusion that authenticism must be false, too. To see this, consider three populations:

O   The original population of some species.
R   A remnant population, whose members are naturally descended from the members of *O*.
D   A de-extinct population, created by synthetic biologists using genetic material recovered from the now-dead members of *O*.

The genesis argument tells us that the resurrection thesis is false, and the resurrection thesis tells us *that D can have approximately as much conservation value as O* (so that in creating *D* we would recoup what we lost when we lost *O*). Thus, the genesis argument tells us, in effect, *that D cannot have approximately as much conservation value as O*. Now, notice that *O* and *R* will share the same conservation value as each other, since the members of *R* are descended from the members of *O* via a normal, natural breeding process. Thus, in implying that D cannot have approximately as much conservation value as *O*, the genesis argument also implies *that D cannot have approximately as much conservation value as R*. This being so, it contradicts authenticism—because authenticism (or, at least, the A2 version thereof) tells us *that D can have approximately as much conservation value as R* (because it tells us that *D* can be an adequate substitute for *R* where *R*'s conservation value is concerned). Thus, the genesis argument isn't just an argument against the resurrection thesis. It is an argument against authenticism as well.

3.4   The Genesis Argument, Refuted

So much for the genesis argument itself. Now, is it sound? We think not. We think its first premise, G1, is false. G1 assumes that de-extinct organisms won't inherit the evolutionary history of the organisms from which their genomic information was copied. This assumption doesn't survive careful scrutiny. To see why not, start by considering the following pair of cases:

Case 1. I visit a natural landscape of huge canyons and coloured rock strata. Awed by its beauty I hope that one day my grandchildren will get to see it too. The landscape is, however, destroyed by a strip-mining venture before my grandchildren are born. But the land is subsequently restored, using terraforming technology, to look just as it used to look: new layers of rock are extruded over it, and canyons are carved through this new rock by terraforming machines. One day my grandchildren get to see the *restored version* of the landscape. Has my hope been fulfilled? Obviously not. The original landscape was ancient, and its ancientness was a major contributor to its aesthetic value. It was this *ancient* landscape I wanted my grandchildren to see. The new, terraformed landscape, of brand new layers of rock and brand new canyons, is, for the very reason of its newness and humans' role in making it, not as good, not as aesthetically valuable, as the ancient landscape it replicates. This is Robert Elliot's argument against the restoration thesis, discussed a few pages ago.

Case 2. I go into the New Zealand bush, hear a kokako—perhaps the world's premiere songbird—sing, and hope that one day my grandchildren will get to enjoy the same exquisite experience. However, the kokako I heard singing eventually dies. So too do all the other kokako that were alive when I heard the kokako sing. But before they all die many of them reproduce, carrying their lineages forward. And before their offspring die in their turn, many of them reproduce themselves too. Eventually, after many generations of kokako have come and gone, my grandchildren walk into the bush and hear a kokako sing. Has my hope been fulfilled? This time the answer is 'yes'. When I hoped that my grandchildren would one day get to share my experience of hearing a wild kokako sing, my hope was not that they would get to hear *numerically the same kokako I heard*—it being perfectly obvious that this particular kokako would be long dead and gone by the time my grandchildren were born. Rather my hope was that my grandchildren would get to hear *some as-yet unconceived* kokako sing, a creature instantiating the same evolved design as the creature I heard, because it was created, via a process of iterated reproduction, from the same stock.

These two cases serve to expose a crucial weakness in the analogy between natural landscapes and natural organisms upon which the genesis argument is based. Where natural landscapes are concerned, a reproduction is, in general, *not as good as the original*. It is not as aesthetically valuable, because it has a different history than the original, and because its history matters greatly to its aesthetics. (The same is obviously true of reproductions of destroyed artworks.) In contrast, where natural organisms are concerned, a reproduction is, in general, *every bit as good as the original*. Every natural organism is, after all, nothing but a product of iterated reproduction; of the repeated *copying* of genes and traits from each generation to the next. It is of the very essence of living organisms that they reproduce, then die. It is not likewise of the essence of natural landscapes, or artworks, that they reproduce, then die. A kokako has parents, who produced it via a replicative act. The Grand Canyon does not.

So much for the disanalogy between natural species and natural landscapes. Now, why is this disanalogy important where de-extinction is concerned and where the soundness of the genesis argument is in question? The answer is simply this: just as natural reproduction involves the copying of genes and traits from one generation to the next, *so too does the type of artificial reproduction that is the cornerstone of a de-extinction project*. De-extinction amounts to *a method for organisms to reproduce*, albeit a method that so happens to involve a great deal of technological and human midwifery, and that also happens to result in the 'offspring' being born into the world years, decades or centuries after the 'parents' have died. It is, so to speak, a form of highly delayed, human-assisted reproduction. One way for a mammoth to reproduce its genes and traits is by successfully mating with another mammoth. Another way for it to reproduce its genes and traits is by dying in such a way that its remains get entombed in the Arctic permafrost and then dug out by us twenty thousand years later to be used as a genetic template for the creation of new, de-extinct mammoths. The latter method leaves a lot more to chance than the former, but can still result in genes and traits being successfully copied.

Let's focus on some particular mammoth that got its dead body frozen in the permafrost. Let's call her 'Beta'. Let's imagine that Beta's genome is now being used by synthetic biologists to create new mammoth-like creatures. Let's call one of the freshly-minted, de-extinct, mammoth-like creatures they have created, 'Gamma'. Since Beta was a natural-born mammoth of the original, ancestral mammoth population, it is obvious that she possessed a full quota of evolutionary history and ecological beauty. According to the genesis argument's first premise, G1, Beta's evolutionary history (and thus her ecological beauty) will not be 'carried over' by the de-extinction process to Gamma. But why not? To see why G1 is problematic, notice that Beta is herself the product of reproduction. Beta had a mother, 'Alpha', as we will name her. Alpha had a full quota of evolutionary history and ecological beauty, and in reproducing her genes and traits to create Beta, she transmitted this history and ecological beauty to Beta. Plainly, therefore, it is possible for an organism's evolutionary history and ecological beauty to be transmitted to future organisms via the copying of genes and traits. It happens all the time.

Indeed, the mechanism by which it happens is not mysterious. By way of understanding the mechanism, consider the example of an architect who designs a house. One group of builders builds a house to the specifications of the architect's design. A second group of builders subsequently (perhaps hundreds of years later) builds a second house, by slavishly copying the design of the first house. Let's suppose that this copying process is reliable in the sense that it is 'counterfactual supporting'. This is to say that if the architect's design had been different than it actually was in any given way, then this difference would have been reflected in the finished structure of the second house. For example, if the architect had set the pitch of the roof at 43° instead of 42°, then the pitch of the second house's roof would accordingly have been 43°, not 42°.

Of course, the second house won't have *all* the same historical properties as the first house. For example, it will have been built at a later date, by a different group of builders. But the two houses will still share at least one important historical property in common: viz., that *of being instantiations of the architect's design*. The second house inherits this property from the first house by direct virtue of its being an accurate reproduction of the first house, created by a high-fidelity, counterfactual-supporting copying process. The operative principle is this:

> *The copying principle*: If A instantiates a design created by X, and B is a high-fidelity, counterfactual-supporting copy of A, then B also instantiates the design created by X.

Now, as for the architect and reproductions of houses, so too for evolution and reproducing organisms. What is the historical property of a living organism that underpins its ecological beauty? On a rough first pass, it is the property *of having been designed by natural selection to efficiently reproduce its genes in such and such an historical environment of evolutionary adaptation*. For example, the historical property Alpha possessed in virtue of which she was ecologically beautiful was the property *of having been designed by natural selection to efficiently reproduce her genes in the megafaunal grazing niche of the mammoth steppe*. Alpha successfully transmitted this property, and thus her ecological beauty, to her daughter, Beta. How did this happen? Well, in just the same way that the architect's design was transmitted from the first house to the second—namely, by invocation of the copying principle. Alpha's genes and traits were of ancient, evolutionary design. These genes and traits were successfully replicated via a high-fidelity, counterfactual-supporting copying process (namely, natural reproduction), to make Beta. And so Beta's genes and traits were also of ancient, evolutionary design. It is as simple as that. Facts about the evolutionary origins of the mammoth's design were passed down the generations by a process of iterated natural reproduction, from a copy, to a copy of the copy, and so on.

Here we reach the crucial point. The very same mechanism that results in Beta inheriting Alpha's evolutionary background and ecological beauty will also result in Gamma inheriting Beta's evolutionary background and ecological beauty. After all, just as Beta was created by the copying of Alpha's genes and traits, so too Gamma is created by the copying of Beta's genes and traits. Admittedly, in the latter case the copying is accomplished by artificial, technological means, and it occurs thousands of years after Beta and all the other mammoths expired. But this is irrelevant where the copying principle is concerned. Beta's genes and traits were of ancient, evolutionary design. Beta's genes and traits have—let us stipulate—been successfully replicated, using a high fidelity, counterfactual-supporting copying process (the copying process used by the scientists in the lab), in order to create Gamma. Thus it follows, by the copying principle, that Gamma's genes and traits are also of ancient, evolutionary design. The genesis argument's first premises, G1, is, for this reason, false. G1 says that the historical properties of the pre-extinction population of organisms will not be carried over by the de-extinction process to the new, de-extinct population. It assumes that there are no historical properties of a thing that will reliably be inherited by a copy of that thing. But the copying principle tells us that this is wrong. There is at least one historical property of a thing that will reliably be inherited by high-fidelity, counterfactual-supporting copies of that thing: namely, the property *of instantiating a design created by so and so.* And it is precisely this kind of property—a property involving the (ancient, evolutionary) origins of a design—that underpins the ecological beauty of organisms.

G1 is false for reasons just explained. Why then does it appear so initially plausible? One reason for its apparent plausibility is because of the analogy with things like natural landscapes and artworks. It is certainly true that a restored version of a destroyed natural landscape, or a replica of a destroyed Picasso, will lack the history, and thus much of the value, of the original. But as we have seen, this analogy is misleading. Natural organisms are always destroyed as a matter of course (i.e., they 'die'), but the historical properties they possess that make them ecologically beautiful are not usually destroyed with them. Rather, these properties (concerning their ancient evolutionary design) are transmitted, through an iterated copying process, into their descendants.

A second reason for G1's apparent plausibility is because of a seeming contradiction in the idea of an animal *artificially created in a biotech lab* having an *ancient evolutionary design*. However, a moment's careful thought shows that any appearance of a contradiction here is illusory. Imagine that Gamma—the newly recreated mammoth—is released into the frozen Siberian tundra. We see her there, kicking back the snow, and tearing up buried grasses with her long prehensile trunk. Her shaggy coat warms her. She is almost invulnerable to frostbite because of a biochemical quirk in her haemoglobin that causes it to convey oxygen efficiently even at very low blood temperatures. Whence came the wonderful design Gamma instantiates—a design so perfectly calibrated for forging a living in these hostile climes? Did it come from the synthetic biologists who made her? No. They built her, but had no part in designing her. Evolution was her architect. The synthetic biologists merely took the design evolution laid down and created a flesh and blood animal faithful to its specifications. They did not design her genes or traits. They copied them, from mammoths in the ancestral population. She is *descended* from these ancestral mammoths, from which her genes were copied, and so she is part of a phylogenetic lineage leading back into the depths of prehistory.

In conclusion, although G1 appears superficially plausible, it is in fact false, and the reasons it appears plausible are easily seen to rest on misunderstandings. Since the genesis argument against authenticism has a false premise, it is unsound and should convince nobody.

3.5   THE EVOLUTIONARY ARGUMENT AGAINST AUTHENTICISM

What we will call the 'evolutionary argument' against authenticism runs as follows:

E1. Organisms created using de-extinction technology will lack the evolutionary history possessed by members of the original members of the extinct species.

E2. If an organism lacks the evolutionary history of the original members of an extinct species, then it is not an authentic member of that species.

C. Therefore, it will never be possible to create authentic new members of an extinct species using de-extinction technology (and so authenticism is false).

The evolutionary argument is very similar to the genesis argument, but attacks authenticism based on an evolutionary conception of species (as referenced by E2), rather than on the basis that de-extinct organisms would lack the conservation value of the originals. Versions of this argument have been defended by numerous critics of de-extinction (Blockstein, 2017; Gunn, 1991, p. 301; Kohl, 2017; Robert, Thévenin, Princé, Sarrazin, & Clavel, 2017, pp. 1026–1027).

We will give the evolutionary argument short shrift, since we have refuted one of its premises already. Its first premise, E1, is identical to the first premise of the genesis argument, G1, and we explained why G1 is false in the previous section. We conclude on this basis that the evolutionary argument is every bit as unsound as the genesis argument.

3.6   THE PHYLOGENETIC ARGUMENT AGAINST AUTHENTICISM

The *phylogenetic species concept* (otherwise known as the *Hennigian species concept*) identifies a species with a segment of the phylogenetic tree (Wheeler & Meier, 2000). As so conceived, a species goes out of existence in either of two ways: first, by speciating (i.e., by dividing into two new species, as represented within the phylogenetic tree by a segment forking to create two new segments); or second, by going extinct (as represented by a segment coming to a dead end).

Ian Smith (2016, pp. 102–104, 2017) points out that, at least on an orthodox understanding of the phylogenetic tree's structure, a segment that has been terminated by an extinction event cannot subsequently reappear again after some intervening gap of time has elapsed ('floating free' of the rest of the tree, as it were). He writes that "Phylogenetic systematicists work with species lineages that are temporally *continuous*" (2016, p. 103), and concludes on this basis that "extirpated species cannot be brought back" (ibid., p. 102). We will call this argument against authenticism the 'phylogenetic argument'.[2] The idea at the heart of the argument is neatly expressed by David Hull, as follows:

If a species evolved which was identical to a species of extinct pterodactyl save origin, it would still be a new, distinct species ... Species are segments of the phylogenetic tree. Once a segment is terminated, it cannot reappear somewhere else in the phylogenetic tree. (1978, p. 349)

We think that the phylogenetic species concept is consistent with authentic de-extinctions being possible, despite what Smith and Hull say.[3] To see why, consider the Alaskan red flat bark beetle (*Cucujus clavipes puniceus*). The larvae of this beetle contain an antifreeze so potent that instead of the larvae eventually freezing when temperatures keep falling, they *vitrify*. That is, the water within them turns into a kind of glass, instead of crystalizing into ice. This enables the larvae to be revived after being kept in cold storage at temperatures down to −100 °C (Sformo et al., 2010). At such temperatures, all metabolic processes cease. Now, let's imagine a scenario in which these beetles were completely extermi-nated save for a collection of their larvae that were being stored in an ametabolic, vitrified state at −100 °C in a cryogenic chamber. If we say that a species is 'extinct' when it has *no living members*, and if we say than an organism is 'living' only *when it is metabolizing*, then in this scenario the Alaskan red flat bark beetle is extinct. Of course, this is an 'extinction' that could easily be reversed by gently thawing the larvae until they start metabolizing again. This would be an example of what, in Chap. 1, we termed a 'cryptobiotic de-extinction'.

What are proponents of the phylogenetic species concept to say about such a scenario? The answer, surely, is that they will happily concede that such cryptobiotic de-extinctions are possible, it being absurd to deny this. (It is, after all, conceivable that such 'extinctions' and subsequent 'de-extinctions' of the beetle have actually happened during historical Alaskan cold-snaps.) When they say that segments of a phylogenetic tree *are termi-nated by extinctions*, they don't mean to claim that a species is necessarily gone forever the moment it no longer has any 'living' (i.e., metabolizing) members. They will, if they are sensible, acknowledge that a species can go into a sort of limbo state (a state of suspended animation) in which it is extinct *in the sense that it has no metabolizing members* but not yet *termi-nally extinct*, where a terminal extinction is the sort of extinction that terminates a phylogenetic segment.

When this idea that a species can exist in limbo state is taken on board then the result is the type of phylogenetic tree depicted in Fig. 3.1.

In Fig. 3.1 each species is represented by a segment of a phylogenetic tree, in accordance with the phylogenetic species concept. But the seg-

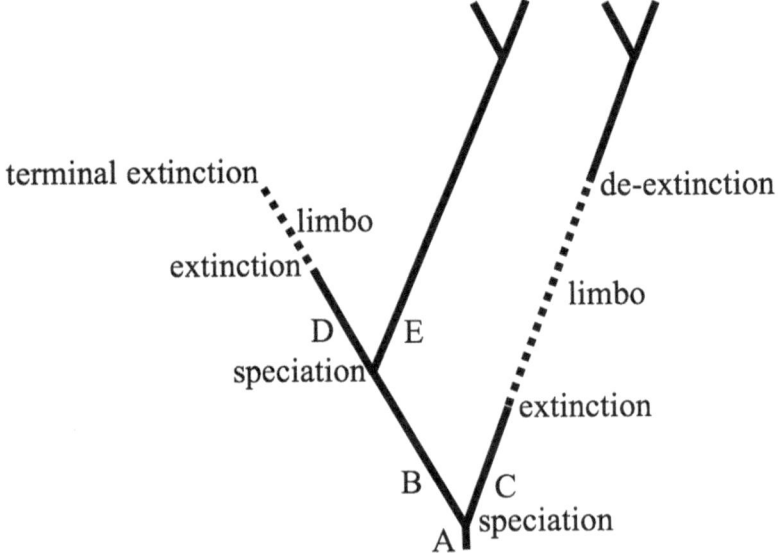

Fig. 3.1 A phylogenetic tree in which species A speciates, producing species B and C, and in which species B then speciates in its turn, producing species D and E. Species C goes extinct, but is subsequently made de-extinct. Species D goes extinct, but is not made de-extinct, and eventually becomes terminally extinct

ments themselves have a sub-structure, with each segment being composed of one or more sub-segments. Each of these sub-segments is either a *solid line* or a *dotted line*. A solid line represents a species during a period when its population size is *greater than zero* (i.e., when *it is not* in limbo). A dotted line represents a species during a period when its population size is zero (i.e., when *it is* in limbo). Transitions from a solid line to a dotted line represent non-terminal extinctions of the species. Transitions the other way, from a dotted line to a solid line, represent de-extinctions. Terminal extinctions are represented by a segment coming to a dead end without branching.

The crucial thing to notice about Fig. 3.1 is that it is consistent *both* with the key idea behind the phylogenetic species concept—namely, the idea that a species is a segment of the phylogenetic tree—*and* with the idea that authentic de-extinctions are possible. To reconcile these *two ideas*, all one must do is sign up to the *third idea* that a species can potentially persist onwards through time in a limbo state even when it has no living members left. We believe that proponents of the phylogenetic species concept have little choice but to sign up to this third idea, because without it they can't explain the obvious fact that de-extinctions of the cryptobiotic variety, at least, are possible. But once they have signed up to this idea in order to explain cryptobiotic de-extinctions, the floodgates are opened. Phylogenetic trees of the type depicted in Fig. 3.1 are consistent not just with cryptobiotic de-extinctions, but with other types of de-extinctions too (gametic de-extinctions, zygotic de-extinctions, clonal de-extinctions, reconstructive de-extinctions, and so on). For example, consider a species that now exists only in the form of a few taxidermied museum specimens. Is this species terminally extinct, or is it in a limbo state from which technology could summon it back? A proponent of the phylogenetic species concept is free to go *either way* in answering this question. Nothing in the phylogenetic species concept itself forbids her from saying that even such a species as this still persists in a limbo state, and is thus still a potential candidate for resurrection.

3.7   THE 100% ARGUMENT AGAINST AUTHENTICISM

According to the paleogeneticist, Beth Shapiro, de-extinct organisms will be inauthentic if they have any tell-tale genetic, physiological or behavioural oddities, however slight, that would mark them apart from the original members of the extinct race. In other words, in her view only a completely exacting de-extinction could be an authentic de-extinction. Shapiro also contends that synthetic biologists are in practice very unlikely to ever be able to create organisms altogether free of such oddities. That is, she thinks completely exacting de-extinctions will never happen. For example, she writes:

> Extinct species are gone forever. We will never bring something back that is 100 percent identical—physiologically, genetically, and behaviorally identical—to a species that is no longer alive. (Shapiro, 2015, p. 10)

> The task ahead is not to make perfect replicas of species that were once alive. [It] is technically not possible to do so and is unlikley ever to be technically possible to do so. (Shapiro, 2015, p. 205)

> Because the phenotype of an organism is the consequence of the interaction between its genotype and the environment in which it develops and lives, even species with cloned nuclear genomes will not be exact copies of the

extinct species on which they are modelled. We should therefore consider de-extinction as a means to create ecological proxies for extinct species. (Shapiro, 2017, p. 996)

The philosopher Jonathan Beever (2017) and conservation biologist Curt Meine (2017) have similarly argued that the products of a de-extinction attempt will be inauthentic due to their failing to share numerous social, environmental and epigenetic properties in common with the original members of a species (down to details like not having the historically correct gut flora). Here are Meine's words:

Technology cannot, and can never, recreate a lost species, because a species is more than its genome; it is the unique expression and evolution of the genome, through a vital population, interacting with and within a unique physical, biological, and social environment, over a unique period of time .... Extinction, in other words, *is* indeed forever. De-extinction, it follows, is a literal impossibility. (p. 13)

We will call this the '100% argument' against authenticism. Here is a formal version:

O1. De-extinction technology will never make it possible to create new organisms that are 100% genetically, physiologically and behaviourally indistinguishable from the members of the original population of an extinct species.

O2. If a de-extinct organism is not 100% genetically, physiologically and behaviourally indistinguishable from members of the original population of the species, then it is not an authentic member of that species.

_____

C.   Therefore, it will never be possible to create authentic new members of an extinct species using de-extinction technology (and so authenticism is false).

Is the 100% argument sound? We think not. Its second premise, O2, is manifestly implausible, and its first premise, O1, is by no means obviously true either.

We will start with O2. Why might O2 be thought true? Shapiro doesn't explain, but in places (e.g., Shapiro, 2015, p. 205) it appears she might have an argument like the following in mind:

If the products of de-extinction don't share all the same genetic, physiological and behavioural properties as the members of the original species, then they *won't be the same thing* as the original species; and if they are not *the same thing* as the original species, then they are *a different thing*: i.e., they are a *different species*, and hence inauthentic. Q.E.D.

However, this argument is fallacious. It assumes that changes through time in the *qualitative properties* of a species signify a change in the *numerical identity* of the species. But it is not true in general that changes in the qualitative properties of a thing signify a change in its numerical identity. For example, if I scratch my spectacles, then my now-scratched spectacles will be qualitatively different from my formerly unscratched spectacles, much to my chagrin. But they will still be numerically the same pair of spectacles. I don't make one pair of spectacles vanish and a new pair appear in their place by scratching them.

As for spectacles, so too for species. Species' living populations don't usually remain static and unchanged over time. To the contrary, mutation, selection, genetic drift and a changing environment are constantly driving shifts in a population's genetic, epigenetic, physiological and behavioural profiles, but we don't declare one species to have passed from the Earth and a new species to have been created in its place every time there is some such change. Instead we regard a species as being able to retain its numerical identity even in the face of sustained, incremental alterations to the qualitative properties of its living population. For example, 10,000 years ago no human beings had blue eyes, the mutation for blue eyes having not yet arisen. Now, on the other hand, blue eyes are commonplace. Thus, our species had, but no longer has, the property *of having no blue-eyed members*. Yet even though this qualitative property of the species has changed, the humans who lived 10,000 years ago were still *Homo sapiens sapiens*, just like us. The species as it was *then* is qualitatively non-identical to the species as it is *now*, but still numerically one and the same species, for all that.

As another example, consider the pigs that were released by sailors onto the uninhabited Auckland Island in 1807. The pigs have thrived on the island as feral animals ever since. Auckland Island is a cold, windswept subantarctic island with a climate and ecology markedly unlike the climate and ecology of Europe, where the pigs came from. This change in environment and diet has almost certainly had numerous subtle effects on the pig's genetics, epigenetics, gut flora and behaviour. But for all these differences between Auckland Island pigs and other pigs, the Auckland Island pigs are still *pigs*. They remain perfectly authentic members of the species *Sus scrofa*. Yes, they were suddenly translocated into a radically alien physical and biological environment, and yes, the phenotype of an organism depends not only on its genes but also on environmental factors. Hence Auckland Island pigs will unquestionably be different in some phenotypic respects from other pigs. Nonetheless, they still enjoy a full measure of pig-hood. (If all the pigs in the rest of the world were to suddenly die of swine-flu, the species would not be extinct so long as the animals of Auckland Island still lived on.)

These examples do not merely serve to undermine the argument *for* O2 outlined above. They also provide us with a powerful argument *against* O2. O2 implies that de-extinct organisms must be *perfect replicas* of the organisms that went extinct if they are to count as being authentic members of the same species. But we have just seen that the members of a species who live at one time are not usually perfect replicas of the members of that same species who lived at earlier times. To the contrary, the genetic, epigenetic, physiological and behavioural properties of a species' living population are in a state of perpetual flux. Hence O2 imposes an *unreasonable* requirement on de-extinction, it being a requirement natural species don't ordinarily satisfy from one year to the next. Why must a mammoth born of de-extinction technology today be *exactly like* the mammoths of 10,000 years ago in order to be authentic, when an elephant born today will be probably be subtly different in all kinds of small genetic, epigenetic and physiological respects from its own parents? And why should we think that a mammoth can be authentic only if it inhabits an environment with exactly the climate and ecology of the Palaeolithic mammoth steppe, when pigs can retain their authenticity even after being translocated into the alien environment of Auckland Island?

If further reasons are wanted for thinking O2 is false, they are easy obtained considering how much an organism can deviate from the norm for its species while still fully belonging to that species. By way of an example, imagine a badly inbred, captive-born, poorly nourished tiger, that is infertile and has spent its whole life in a cramped cage. This unfortunate tiger will differ sharply from wild tigers in genetic, physiological and behavioural respects, yet it is still *a tiger*, not a creature of a different species. It is not a mere pseudo-tiger. If its continued existence lacks much in the way of conservation value then this will be because of the *kind* of tiger it is—an infertile, mentally disturbed tiger with bad genetics—not because it is *not* a tiger. It is *a poor example of a tiger*, but, therefore, a tiger for all that. Despite its genetic, physiological and behavioural inadequacies, it still doesn't cross the line separating tigers from non-tigers. By the same token, if tigers went extinct and a de-extinction project aimed at resurrecting the species produced only such infertile, deranged, genetically compromised animals, then we could not on this basis conclude that it had failed to produce authentic tigers. The threshold *for authenticity* is not to be confused with the much more demanding threshold *for being a fully fit, normal and healthy specimen of a species.*

That Shapiro's endorsement of O2 is a mistake becomes even clearer in light of the fact that O2 subtly contradicts another claim she endorses: namely, the claim that *facilitated adaptation* is a legitimate conservation technique. Shapiro approvingly discusses two examples of facilitated adaptation. The first example is the idea, examined in Chap. 2, of re-engineering Asian elephants for cold tolerance so that they can escape human population pressure in the tropics. The second example is an already completed project, mentioned in Chap. 1, to paste a wheat gene into the genome of the American chestnut to give the chestnut resistance to the devastating chestnut blight fungus. These would be sound methods of saving the Asian elephant and the American chestnut from extinction only if the cold-tolerant elephants and blight-resistant chestnut trees thus produced *still counted as being authentic members of the original species,* despite their altered genetics, physiology and behaviour. After all, one obviously wouldn't save Asian elephants from extinction by turning them into creatures that weren't Asian elephants anymore—and likewise for the chestnut trees. For this reason, Shapiro is herself tacitly committed to the idea *that a species can retain its authenticity even in the face of sudden and substantial, human-caused changes to its genetics, physiology and behaviour.* But this being so, why does she insist that a resurrected species can be authentic only if no such changes—not even very small ones—have taken place? There is a double standard at play.

Next, O1. We will keep our remarks on O1 brief, since in refuting O2 we have already shown the 100% argument to be unsound. Shapiro's arguments for O1 are based on the limitations of cloning technology, including: (1) the fact that the clones will have different mitochondrial genetics than the originals (since their mitochondria will come from the donor of the ovum, not from the extinct species); (2) effects on a foetus' development caused by its developing in the uterus of a surrogate mother of a different species; (3) effects on a clone's behaviour caused by its being raised by a surrogate species; and (4) effects on a clone's phenotype and epigenetics caused by differences in environmental stressors and diet (Shapiro, 2017, pp. 1000–1001). The problem with this argument is that issues like (1)—(4) needn't arise in relation to *all* de-extinction methods and *all* target species. To see this, consider the Pacific oyster (*Crassostrea gigas*). These oysters are unusual among animals that fertilize externally (including many species of frogs, fish and marine invertebrates) in the respect that procedures already exist not only for cryogenically preserving their sperm (which is easy for most species) but also for cryogenically preserving their oocytes (Tervit et al., 2005). Now imagine that the Pacific oyster species was to go suddenly extinct, but not before a genetically diverse sample of its sperm and oocytes had been cryogenically stored. Could we resurrect the species again? Yes. It would be easy. It would simply be a matter of unfreezing the sperm and oocytes, mixing them together in seawater, and letting nature take its course. There is in this case no need for cloning, which avoids the 'wrong mitochondria' problem that Shapiro mentions. Furthermore, since oysters fertilize externally rather than internally, and since their larvae develop free-floating in the ocean, there is no need for a surrogate mother, which at one stroke avoids all the problems Shapiro mentions in connection to surrogacy. Finally, since Pacific oysters don't transmit behavioural patterns from one generation to the next via learning, there will also be no 'cultural' behaviour to be lost. Let's imagine that new Pacific oysters are allowed to develop in a contained volume of seawater that is indistinguishable from the seawater that the pre-extinction oysters typically developed in. (Maybe we have carefully arranged for the volume of seawater in question to have exactly the right pH and temperature, and to contain exactly the right phytoplankton.) This being so, would there be any strong reasons for thinking that the resurrected Pacific oysters must differ in some genetic, physiological or behavioural respect from the members of the original, extinct oyster population? It would appear not. Certainly, if there are such reasons then Shapiro hasn't given them. And so it appears that the de-extinction of the Pacific oyster, or of any species akin to the oyster, could provide a counterexample to O1.[4]

In summary, the 100% argument is unpersuasive. O2 is patently implausible. O1 is open to challenge, too.

3.8   The Argument from the Definition of 'Extinction' Against Authenticism

Allister Gunn (1991) argues that it is part of the very concept of 'extinction' that extinctions, properly so-called, are *irreversible*. As evidence, he cites the *Shorter Oxford Dictionary*'s definition of 'extinction'—*a coming to an end or dying out*—and notes that "this definition sounds permanent enough" (p. 299). No species that is 'extinct' in *this* sense of the word will ever be successfully resurrected, for if it *were to be* resurrected then it wouldn't have truly gone extinct in the first place. In short, if by an 'authentic de-extinction' we mean the reversal of such an 'extinction' then authentic de-extinctions are logically impossible, and so authenticism is false. Q.E.D.

This argument has been convincingly rebutted by Helena Siipi (2014), who points out that even if we accept Gunn's characterization of the extinction concept (which we need not, because we can instead define extinction in terms of a species' population size having fallen to zero), then at most it follows that we should not *describe* a species as having gone extinct if its numbers fell to zero but were subsequently boosted above zero again. It doesn't follow that it is impossible to boost the numbers of a species in this way, and so it doesn't follow that the sorts of events that we have been calling 'authentic de-extinctions' in this book are impossible. (Gunn might just as well have argued that authentic de-extinctions are logically impossible by redefining 'authentic de-extinction' to mean 'square circle'. Yes, square circles are logically impossible, but that's changing the subject!)

3.9   The Argument from Taxonomy Against Authenticism

Gunn also argues against authenticism based on the species-naming practices of taxonomists. He notes that the Australian red back spider and American black widow are classified as being one species, *Lactrodactus mactans*, while the very similar New Zealand katipo spider is classified as another, *Lactrodactus scelio*, and contends that "If such similar products of non-human-assisted evolution are separate species, then surely so are the products of human genetic manipulation" (1991, p. 300).

Perhaps Gunn is right that taxonomists will invent new scientific names for de-extinct organisms. For example, perhaps de-extinct woolly mammoths will be classified as being members of a new, *Mammuthus primigenius-de-extinctus* species instead of the old *Mammuthus primigenius* species. But such facts about taxonomic conventions are irrelevant to the truth or falsity of authenticism. Where authenticism is concerned, what is important is just the *conservation value* of the de-extinct organisms. If the creatures produced in an attempt to resurrect the woolly mammoths would share a conservation value similar to that of any putatively existing remnant population of woolly mammoths, then they are authentic. Otherwise, they aren't. The label taxonomists choose to pin on the creatures is neither here nor there.

3.10   THE POPULATION-BOOSTING ARGUMENT
FOR AUTHENTICISM

That completes our survey of prominent arguments *against* authenticism. We believe none of them are persuasively forceful. We turn now to two arguments *for* authenticism, both of which we believe *are* persuasively forceful.

The first argument—the 'population-boosting argument', as we will call it—is inspired by what we believe to be a lurking contradiction in the position advocated by Beth Shapiro (2015). Although Shapiro is an anti-authenticist who doesn't think de-extinction techniques can be used to create authentic new members of extinct species, she does think these techniques can be used to *boost the populations of extant species.* One such population-boosting technique she mentions is interspecies cloning. This involves using somatic cells from an endangered species and the surrogacy services of some other related species to create new members of the endangered species. It has already been used to create mouflon using the surrogacy services of sheep (Loi et al., 2001), and to create guar using the surrogacy services of cattle (Srirattana et al., 2012; Vogel, 2001). Another technique Shapiro mentions is iPGCT (described in Chap. 2) which has been used to produce chicken offspring from ducks (Liu et al., 2011) and bustard offspring from chickens (Wernery et al., 2011). Shapiro explains how the latter technique can be used to produce "a pure-bred rare-breed chicken that hatches from the egg laid by a common chicken", so as to "boost the population size of rare or endangered chicken breeds" (2015, p. 156).

Shapiro is supportive of using these methods to boost the populations of endangered species. But notice that it is only a small step from doing this to using the same methods to resurrect an extinct species. The small step in question simply consists of the species going extinct. 'Boosting' a species' population size is *the same thing* as de-extinction in the special case that the species' present population size happens to be zero. If a species' population size has fallen all the way to zero and if then we use a population-boosting technique to 'boost' its population size by creating new members of the species, then we will thereby have undone the extinction of the species.

There appear to be three options open to Shapiro at this juncture.

Option 1 is to renounce her view that techniques like interspecies cloning and interspecies primordial germ cell transplantation can be used to create authentic new members of an extant species. This would entail claiming that none of the various animals that have been produced using these methods so far—including mouflon, guar, chickens, and bustards—are authentic members of the species they appear to belong to. It would entail insisting that these animals are, in truth, nothing better than mere pseudo-mouflon, pseudo-guar, pseudo-chickens and pseudo-bustards. Option 1 appears desperately unattractive, especially in cases where the animals in question are physiologically normal, fertile, and fully capable of breeding with animals that have not been created using these technologies. It appears all the more unattractive when one considers that human babies could potentially be created by similar methods (perhaps using a chimpanzee surrogate mother). We would presumably be loath to classify people created in this way as being mere *pseudo-Homo-sapiens*.

Option 2 is to stand by the view that these methods can be used to boost the population size of a species, but while also insisting *that they can't be used to 'boost' it from a starting point of zero*. Again, however, this is a position that would be difficult to sustain. To see why, consider an example involving mouflon. No live mouflon are required in order to create a mouflon using interspecies cloning. All that's needed are cryogenically preserved mouflon cells. Given a supply of these cells one can set about creating mouflon using cloning and the surrogacy services of sheep. Imagine that at *almost* the same moment when an animal created in this way was being born, all the living mouflon in the world were killed (say, by simultaneous lightning bolts, or by a gas explosion in a mouflon breeding centre). In this case, would the new-born animal be an *authentic mouflon* or a *pseudo-mouflon*? If Shapiro was to embrace Option 2 then she would need to say that the answer depends on precisely *when* all the other mouflon met their sudden deaths. If they were killed a moment *after* the birth of the animal (so that there was a short period of overlap) *then the animal would be a mouflon*. In this case the potential would still exist to restore the mouflon species to a thriving state by creating many more mouflon in the same way. On the other hand, if they were killed a moment *before* the birth of the animal, creating a brief period of time in which there were no living mouflon-like animals on the planet, *then the new-born animal would only be a pseudo-mouflon* and the species would have been lost forever. In short, under Option 2 Shapiro would need to accept that the conservation value of the new-born animal (is it the potential saviour of the species, or not?) would be completely determined by facts about *the precise times of death* of all the other animals. We take this to be manifestly absurd. (Who cares whether they all died a couple of seconds before the animal was born, or a couple of seconds afterwards?)[5]

Option 3 is for Shapiro to accept that methods like interspecies cloning and interspecies primordial germ cell transplantation can in fact be used to boost a species' population size from a starting point of zero. This means renouncing her anti-authenticist position and conceding that authentic de-extinctions are possible.

Since Options 1 and 2 both appear desperately unattractive, we believe Shapiro has little choice but to embrace Option 3. And as for Shapiro, so too for other anti-authenticists.

3.11   A Theory of Authenticity for De-extinct Organisms

Our second argument for authenticism is based on a simple theory of authenticity, the 'genotype/phenotype theory' (or G/P theory), which we will now describe. We believe this theory possesses considerable *prima facie* plausibility. It entails authentic de-extinctions are possible. Hence it puts the ball firmly in the anti-authenticists' court. It is their burden to say what's wrong with the theory—a burden which, we believe, they will have difficulty discharging. If they have no good reasons to reject the theory then they have no good reasons for rejecting authenticism.

The genotype/phenotype theory is based on various themes that have emerged already during the course of this chapter:

1. De-extinct organisms are authentic if and only if we would, by creating them, recoup the great bulk of the conservation value that was lost when the original population went extinct.

2. The conservation value of organisms partly inheres in their ecological beauty, and thus in their evolutionary history.
3. De-extinct organisms will inherit the evolutionary history, and thus the ecological beauty, of the original members of a species if their genes are high fidelity, counterfactual-supporting copies of the originals' genes.
4. In order for de-extinct organisms to be authentic they don't need to be perfectly physically indistinguishable from the originals in all genetic, physiological and behavioural respects.

These ideas can be distilled down into the following theory of authenticity for de-extinct organisms, which we submit to the reader's consideration:

The genotype/phenotype theory (G/P theory). A de-extinct organism counts as being an authentic member of an extinct species, *S*, just to the degree that: (i) its genes are high fidelity, counterfactual supporting copies of ancestral *S* genes; and (ii) these genes are being phenotypically expressed in the de-extinct organism in the same way they were expressed in organisms of the ancestral population.

Crucially, the G/P theory entails that authenticity *comes in degrees*, with an organism's degree of authenticity depending on how closely its genes and phenotypic traits match those of its historical precursors. Some examples will illustrate this idea.

Example 1. A de-extinct creature has a genome that perfectly matches that of a long-dead animal, except for one tiny copying mistake. It has an AAG codon instead of an AAC codon at some position in a protein-coding portion of its genome. Luckily it so happens that AAG and AAC both code for the same amino acid, lysine. This genetic mistake therefore has no phenotypic upshots whatsoever. It is a genotypic difference that makes no phenotypic difference. This being so, the G/P theory implies that the mistake will detract from the de-extinct animal's authenticity to an utterly negligible extent.

Example 2. As for Example 1, but the de-extinct creature has an AAA codon instead of an AAC codon at some position in a protein-coding portion of its genome. AAA codes for asparagine, not lysine. This difference in amino acid has, let us suppose, major ramifications for the structure and function of the resulting protein, which in its turn has major, snowballing, downstream phenotypic effects because it prevents many other genes from being expressed in their historically normal ways (say, because the protein is a hormone that is critical to some important developmental process). This being so, the G/P theory implies that the animal will have little authenticity, notwithstanding its genome being an almost perfect replica of the ancestral genome.

Example 3. A de-extinct creature has several improperly copied protein-coding or regulatory genes, but these copying errors have only relatively minor and contained phenotypic effects. (Perhaps they merely affect the animal's eye colour, for example.) The G/P theory implies of such a creature that it is discernibly less than perfectly authentic but quite highly authentic nonetheless.

Example 4. A de-extinct creature has a completely error-free genome. (Its genome is, perhaps, a perfect copy of that of some long-dead animal down to the very last base-pair.) However, a variety of extra-genetic factors cause many of its genes to be phenotypically expressed differently than they were historically. These factors might include such things as differences in epigenetic markers, differences in the gene's cellular environment, differences in the creature's nutrition, differences in the creature's social environment, and so on. This being so, the organism's physiology and behaviour differ sharply from that of its historical antecedents. (Its genotype is right but its phenotype is badly awry.) The G/P theory implies of such a creature that it will have low authenticity. (Thus, the G/P theory doesn't embody the fallacy of genetic determinism.)

Example 5. The aurochs is resurrected, but the animals are found to be so extraordinarily dangerous, unpredictable and evil-tempered that it is politically inexpedient to use them to rewild those parts of Europe commonly frequented by hikers and day-trippers. For this reason, a new breed of *placid aurochs* is created, which differ from the ancestral aurochs only in being disinclined to charge, gore, and toss people without provocation. In all other genetic, physiological and behavioural respects the placid aurochs are just like the aurochs of 10,000 years ago. The G/P theory implies that the placid aurochs will be highly authentic but discernibly less than per-fectly authentic.

Example 6. A new, transgenic strain of the American chestnut is planted-out across the tree's former range in the forests of the Eastern USA. It is genetically and phenotypically indistinguishable from the American chest-nut of old except that its genome contains an extra gene, derived from the wheat plant, that confers resistance to chestnut blight. The G/P theory implies that these trees will have a very high degree of authenticity but that they won't quite be perfectly authentic. (The theory has similar impli-cations with respect to other instances of facilitated adaptation.)

Example 7. The passenger pigeon undergoes a *discriminating de-extinction*—which is to say that deleterious genes possessed by ancestral passenger pigeons are deliberately excluded from the genomes of the newly created birds. As a result the new birds have an unnaturally low mutation load, making them relatively invulnerable to inbreeding depres-sion and giving them a reproductive advantage as they battle to re-establish themselves in a greatly altered environment. Birds created in this way will have genes that have been accurately copied from ancestral birds, and that are phenotypically expressed in the new birds just as they would have been in the ancestral birds. They will differ from past passenger pigeons only in having particularly good, handpicked *sets* of genes and traits (because of deleterious genes having been omitted). Thus, the condition laid down by the G/P theory is satisfied in full, and the G/P theory implies that they will be fully authentic.

   Is this a problem for the G/P theory? Should these birds, with their unnaturally high reproductive fitness, instead be classified as inauthentic? We think not. If one imagines the genes in a species' genepool being randomly shuffled and then dealt out to create a set of all possible organ-isms of that species, then the resulting bell curve of reproductive fitness will have outliers, in the form of organisms that are either very well endowed, or very poorly endowed, in the reproductive stakes. In our view, no matter how much of an outlier an organism is, it will still be an authen-tic member of the species provided all its genes come from the species' genepool. We wouldn't consider a naturally bred organism to be inau-thentic just because it was exceptionally blessed in its genetics, and so we shouldn't consider de-extinct organisms to be inauthentic on this basis either.

Example 8. The New Zealand huia is resurrected and re-established in the wild. The de-extinct birds have no extant population of huia to learn the local dialect of huia song from, and so they gradually invent a new song of their own. This new song is no less delightful to the human ear than the songs the huia sung long ago, before the species was hunted to extinction, but it consists of a different series of notes than any of the old songs did. There is, let's imagine, no clear fact of the matter as to which particular song the huia's genes caused the birds to sing in the past, for the songs were highly mutable, and varied greatly from time to time and place to place. Since there is no such fact of the matter, the G/P theory implies that the resurrected huia will be fully authentic. The huia's genes merely specify that *some* song be sung, not what notes it must consist of, and so—since the resurrected huia do indeed sing a song—the conditions laid down by the G/P theory are met in full. Of course, the song itself, being new, will not be authentically the same as any of the old songs. We will have authentically resurrected the huia species, but not the huia's old songs.[6]

Example 9. A number of woolly mammoth genes for cold tolerance are inserted into the genome of the Asian elephant, and the resulting 'mammophants' are released into northern Siberia. Will these mammophants count as being authentic woolly mammoths according to the G/P theory? No, for only a tiny proportion of their genes will be high-fidelity, counterfactual-supporting copies of woolly mammoth genes. Will they instead count as being authentic Asian elephants according to the G/P theory? Yes. Or at least, they will count as being authentic to a *high*, albeit *non-maximal*, degree. After all, the vast majority of their genes will be high-fidelity, counterfactual-supporting copies of Asian elephant genes, and these genes will (let's suppose) be phenotypically expressed in the mammophants almost exactly as they were in ancestral Asian elephants— the mammophants differing from Asian elephants hardly at all except for being shaggy, having small ears, having thick subcutaneous fat, and having cold-efficient haemoglobin.

So much for the G/P theory itself. We will now explain why it implies that authentic de-extinctions are possible, using three examples:

Example 1: a cryptobiotic de-extinction. Imagine that the Alaskan red flat bark beetle population was reduced down to just a handful of larvae, that were vitrified at −100 °C, causing their metabolism to stop, before being reanimated. In this case the pre-extinction population would consist of the beetle larvae before they were vitrified, and the de-extinct population would consists of the same larvae after they had been brought back to life. The G/P theory will classify these de-extinct organisms as being fully authentic members of the 'extinct' species for obvious reasons: their genes will be the same as those in the pre-extinction population (for they will have been accurately 'copied' from the pre-extinction population by simple virtue of the DNA strands remaining intact during the vitrification process), and these genes will be phenotypically expressed within the reanimated larvae in the historically normal way.

Example 2: a gametic de-extinction. Imagine that the Pacific oyster goes extinct after its sperm and ova have been cryobanked. We then resurrect it by mixing the thawed sperm and ova and letting the resulting zygotes develop in seawater. The new oysters thereby produced will have the same genes as the pre-extinction oysters, and these genes will (all going well) be expressed in the historically normal way—and so they will be highly authentic Pacific oysters according to the G/P theory.

Admittedly, neither the cryptobiotic de-extinction of the Alaskan red flat bark beetle nor the gametic de-extinction of the Pacific oyster would be terribly impressive as species de-extinctions go. They are at the trivial end of the de-extinction spectrum. Are less trivial authentic de-extinctions also possible according to the G/P theory? This brings us to the final example.

Example 3: the clonal de-extinction of a mammal. Consider de-extinction's one major 'proof of concept' to date—namely, the cloning of Celia. Recall from Chap. 1 that Celia's clone had the nuclear genome of a bucardo (obtained from Celia), but the mitochondrial genome of a goat (obtained from a goat who donated an ovum). The mitochondrial genome, with its mere thirteen protein coding genes (Pietro, Maria, GianFranco, & Giuseppe, 2003), is tiny in comparison to the nuclear genome, with its approximately twenty thousand protein coding genes (Dong et al., 2013). Moreover, since the bucardo is a close relative of the goat, it can be presumed that goat mitochondrial genes are a very close approximation of bucardo mitochondrial genes and that they will therefore 'play well' with bucardo nuclear genes in the biochemical environment of the cell. Evidence for this conclusion is provided by the fact that Celia's clone was a physiologically normal bucardo kid except for the lung abnormality that killed it within a few minutes (Folch et al., 2009). Such lung abnormalities were common in clones produced in the early 2000s, when the cloning of mammals was still in its earliest infancy. However, the technology is now much more reliable (Zimmer, 2013). This being so there is presently no technological obstacle to our being able to create a healthy clone of Celia—an animal that would be a physiologically normal bucardo kid in all respects. The overwhelming majority of this kid's genes would be high fidelity copies of bucardo genes, with the only exceptions being its mitochondrial genes. These mitochondrial genes would be similar enough to bucardo mitochondrial genes not to interfere significantly in phenotypic expression of all the kid's other genes. And so the G/P theory entails that this kid would be a highly (if not perfectly) authentic bucardo.

Because the G/P theory entails that authentic de-extinctions are possible even using today's technology, anti-authenticists must repudiate it. As we say, we don't think this will be easy for them to do, since the G/P theory appears very plausible. There is, however, one obvious line of attack they might follow. They might challenge the idea at the heart of the G/P theory—namely, that authenticity *comes in degrees*. Surely—the anti-authenticist might argue—an animal is either an Asian elephant, or not. It can't be *mostly*, but *not completely*, an Asian elephant.

By way of seeing why this objection falls flat, imagine that someone modifies just one gene in the Asian elephant genome, a gene with a single, very localized effect. For example, it might be a gene that only affects eye colour. The resulting elephant has strangely coloured eyes, but in all other respects it is a perfectly genetically and phenotypically normal elephant. With respect to its eye-colour gene, it is 0% authentic. With respect to all of its other genes and traits, it is 100% authentic. The eye-colour gene has been designed by us. It has not been sculpted by millions of years of natural selection. But the rest of the elephant's genome has been so sculpted. What then are we to say about the complete package, the strange-eyed elephant as a whole? We believe it makes perfect sense to say that the elephant as a whole is mostly, but not quite completely, an authentic Asian elephant. It is, to put a rough figure on it, 99.999% authentic. Eye colour aside, it is as authentic as could be. From a conservationist perspective, Asian elephants of this kind would not be as valuable as 100% authentic Asian elephants, with ordinary eyes. But they would be much, much better than no Asian elephants at all. If, through some bizarre turn of events, the world was left with only these strange-eyed elephants, then that would at least be far preferable to it being left without any elephants. Some ancient genes for elephant eye-colour would have been lost, but at least the species' other genes and traits, with their associated outward beauty, functional beauty and ecological beauty, would have survived.

NOTES

1. This section and the next are a reworked version of Campbell (2017).
2. Jebari (2016, p. 218) endorses a similar argument, but restricts its scope to the products of reconstructive de-extinctions.
3. We believe the phylogenetic argument against authenticism can also be challenged on the basis that the phylogenetic species concept is the wrong species concept to be using when assessing the authenticity of de-extinct organisms. However, we won't defend this claim here.
4. Cryptobiotic de-extinctions pose a similar threat to O1. Shapiro could respond to such criticisms of O1 by restricting the scope of the 100% argument to cover only non-cryptobiotic species that fertilize internally, but this wouldn't address our criticisms of O2.
5. It might be claimed that it is not *the moment of birth* that matters, but the moment when some other milestone is reached (say, the moment of conception). But this doesn't matter to the argument, since, for whatever moment, m, is nominated, the thought experiment can be modified to have all the mouflon die either shortly before or shortly after m.
6. We have heard it argued that if it is not possible to resurrect the huia *complete with their old songs* then it is not worth resurrecting the huia at all. This is the fallacy of letting the perfect be the enemy of the good.

REFERENCES

Beever, J. (2017). The ontology of species: Commentary on Kasperbauer's "should we bring back the passenger pigeon? The ethics of de-extinction". *Ethics, Policy & Environment, 20*(1), 18–20. https://doi.org/10.1080/2155 0085.2017.1291825

Blockstein, D. E. (2017). We can't bring back the passenger pigeon: The ethics of deception around de-extinction. *Ethics, Policy & Environment, 20*(1), 33–37. https://doi.org/10.1080/21550085.2017.1291826

Campbell, D. I. (2017). On the authenticity of de-extinct organisms, and the genesis argument. *Animal Studies Journal, 6*(1), 61–79.

Cohen, S. (2014). The ethics of de-extinction. *NanoEthics, 8*(2), 165–178.

Dong, Y., Xie, M., Jiang, Y., Xiao, N., Du, X., Zhang, W., … Wang, W. (2013). Sequencing and automated whole-genome optical mapping of the genome of a domestic goat (Capra hircus). *Nature Biotechnology, 31*(2), 135–141.

Elliot, R. (1982). Faking nature. *Inquiry, 25*(1), 81–93.

Folch, J., Cocero, M. J., Chesné, P., Alabart, J. L., Domínguez, V., Cognié, Y., … Vignon, X. (2009). First birth of an animal from an extinct subspecies (Capra pyrenaica pyrenaica) by cloning. *Theriogenology, 71*(6), 1026–1034. https://doi.org/10.1016/j.theriogenology.2008.11.005

Gunn, A. S. (1991). The restoration of species and natural environments. *Environmental Ethics, 13*(4), 291–310.

Hausdorf, B. (2011). Progress toward a general species concept. *Evolution, 65*(4), 423–431.

Hull, D. L. (1978). A matter of individuality. *Philosophy of Science, 45*(3), 335–360.

Jebari, K. (2016). Should extinction be forever? *Philosophy and Technology, 29*(3), 211–222.

Kohl, P. (2017). Using de-extinction to create extinct species proxies; Natural history not included. *Ethics, Policy & Environment, 20*(1), 15–17. https://doi.org/10.1080/21550085.2017.1291832

Liu, C., Khazanehdari, K. A., Baskar, V., Saleem, S., Kinne, J., Wernery, U., & Chang, I.-K. (2011). Production of chicken progeny (Gallus gallus domesticus) from interspecies germline chimeric duck (Anas domesticus) by primordial germ cell transfer. *Biology of Reproduction, 86*(4), Art. 101, 1–8. https://doi.org/10.1095/biolreprod.111.094409

Loi, P., Ptak, G., Barboni, B., Fulka, J., Cappai, P., & Clinton, M. (2001). Genetic rescue of an endangered mammal by cross-species nuclear transfer using post-mortem somatic cells. *Nature Biotechnology, 19*(10), 962–964. https://doi.org/10.1038/nbt1001-962

McKibben, B. (1989). *The end of nature.* New York: Random House.

Meine, C. (2017). De-extinction and the community of being. *Hastings Center Report, 47*, S9–S17. https://doi.org/10.1002/hast.746

Minteer, B. A. (2015). The perils of de-extinction. *Minding Nature, 8*(1), 11–17.

Pietro, P., Maria, F., GianFranco, G., & Giuseppe, E. (2003). The complete nucleotide sequence of goat (Capra hircus) mitochondrial genome. *DNA Sequence, 14*(3), 199–203. https://doi.org/10.1080/1042517031000089487

Robert, A., Thévenin, C., Princé, K., Sarrazin, F., & Clavel, J. (2017). De-extinction and evolution. *Functional Ecology, 31*(5), 1021–1031. https://doi.org/10.1111/1365-2435.12723

Sandler, R. (2013). The ethics of reviving long extinct species. *Conservation Biology, 28*(2), 354–360.

Sformo, T., Walters, K., Jeannet, K., Wowk, B., Fahy, G. M., Barnes, B. M., & Duman, J. G. (2010). Deep supercooling, vitrification and limited survival to −100°C in the Alaskan beetle Cucujus clavipes puniceus (Coleoptera: Cucujidae) larvae. *Journal of Experimental Biology, 213*(3), 502–509. https://doi.org/10.1242/jeb.035758

Shapiro, B. (2015). *How to clone a mammoth: The science of de-extinction.* Princeton: Princeton University Press.

Shapiro, B. (2017). Pathways to de-extinction: How close can we get to resurrection of an extinct species? *Functional Ecology*, *31*(5), 996–1002. https://doi.org/10.1111/1365-2435.12705

Siipi, H. (2014). The authenticity of animals. In M. Oksanen & H. Siipi (Eds.), *The ethics of animal re-creation and modification: Reviving, rewilding, restoring.* London: Palgrave Macmillan.

Smith, I. A. (2016). *The intrinsic value of endangered species.* New York: Routledge.

Smith, I. A. (2017). De-extinction and the flourishing of species. *Ethics, Policy & Environment*, *20*(1), 38–40. https://doi.org/10.1080/21550085.2017.1291834

Srirattana, K., Imsoonthornruksa, S., Laowtammathron, C., Sangmalee, A., Tunwattana, W., Thongprapai, T., ... Parnpai, R. (2012). Full-term development of gaur–bovine interspecies somatic cell nuclear transfer embryos: Effect of trichostatin A treatment. *Cellular Reprogramming*, *14*(3), 248–257. https://doi.org/10.1089/cell.2011.0099

Tervit, H. R., Adams, S. L., Roberts, R. D., McGowan, L. T., Pugh, P. A., Smith, J. F., & Janke, A. R. (2005). Successful cryopreservation of Pacific oyster (Crassostrea gigas) oocytes. *Cryobiology*, *51*(2), 142–151. https://doi.org/10.1016/j.cryobiol.2005.06.001

Vogel, G. (2001). Endangered species. Cloned gaur a short-lived success. *Science*, *291*, 409.

Wernery, U., Liu, C., Baskar, V., Guerineche, Z., Khazanehdari, K. A., Saleem, S., ... Chang, I.-K. (2011). Primordial germ cell-mediated chimera technology produces viable pure-line houbara bustard offspring: Potential for repopulating an endangered species. *PLoS One*, *5*(12), 1–8. https://doi.org/10.1371/journal.pone.0015824

Wheeler, Q. D., & Meier, R. (Eds.). (2000). *Species concepts and phylogenetic theory: A debate.* New York: Columbia University Press.

Wilkins, J. (2009). *Species: A history of the idea.* Berkeley, CA: University of California Press.

Zimmer, C. (2013, April). Bringing them back to life. *National Geographic*, 28–43.

CHAPTER 4

Ethical Arguments For and Against De-extinction

Abstract This chapter surveys and critically evaluates all the main arguments both for and against de-extinction. It presents a qualified defence of the claim that conservationists should embrace de-extinction. It ends with a list of do's and don'ts for conservationist de-extinction projects.

Keywords De-extinction • Ethics • Biodiversity • Justice • Moral hazard

4.1   INTRODUCTION

Should conservationists deem de-extinction to be good or bad, ethical or unethical? Is de-extinction something conservationists should embrace, or not? This 'Ethical Question' could barely be any more controversial than it is. At one extreme, some conservationists think we have an absolute moral duty to resurrect species we have exterminated if we can (Archer, 2013). At the other, some consider de-extinction to be an egregious violation of nature that would double-down on the original environmental sin of extirpating a species. (Many supporters of Deep Ecology fall in this latter camp.) This chapter will survey and critically evaluate the main arguments on both sides of the de-extinction debate.

Before we begin, we will say a few words by way of clarifying the terms of the dispute. The Ethical Question, as we have defined it, takes the following form:

Ethical Question.   Should conservationists judge de-extinction to be ethical? (Should they embrace it, or not?)

The Ethical Question is not to be confused with Q:

Q.   Is de-extinction ethical?

The two questions are different because conservationists are wedded to certain doctrines, such as D1 and D2:

D1.   Conservationism is an ethical enterprise.
D2.   Biodiversity is worth saving and protecting.

Conservationist's acceptance of these doctrines imposes logical constraints on what they can say about the ethics of de-extinction. Below, when we are evaluating the persuasive power of arguments *for* de-extinction, we will sometimes point out that conservationists have little choice but to *accept* a premise of an argument, because denying it would be tantamount to denying some fundamental conservationist doctrine like D1 or D2. We will consider this a strong reason to think the argument is forceful. Similarly, when we are evaluating the persuasive power of arguments *against* de-extinction, we will sometimes point out that conservationists have little choice but to *reject* the premise of an argument, because the premise could be accepted only on pain of repudiating some doctrine like D1 or D2. We will consider this a strong reason to think the argument is *not* forceful. Our concern throughout will be with whether conservationists should embrace de-extinction, not with what non-conservationists should think of it.

Overview: The arguments for and against de-extinction to be considered in this chapter are these:

For:

- *The biodiversity restoration argument*: de-extinction is ethical because it is a means to the end of restoring lost biodiversity.

- *The justice argument*: de-extinction is ethical because justice demands that we resurrect species we have exterminated.
- *The ecological restoration argument*: de-extinction is ethical because resurrecting extinct keystone species is a means to the end of restoring lost ecosystems.
- *The smart politics argument*: de-extinction is ethical because it is a means to the end of securing greater public backing for conservation.
- *The argument from coolness*: de-extinction is ethical because it would be 'cool'.
- *The balance restoration argument*: de-extinction is ethical as a means to the end of restoring nature's balance, which we have disrupted.

Against:

- *The resource allocation argument*: de-extinction is unethical because it would involve the misallocation of precious conservation resources.
- *The rewilding argument*: de-extinction is unethical because in practice it will be technically impossible to re-establish the resurrected species in the wild.
- *The moral hazard argument*: de-extinction is unethical because it will enable extinctions to be excused on the basis that extinct species can always be resurrected again at a later date.
- *The hubris argument*: de-extinction is unethical because it is risky and something will go wrong.
- *The playing god argument*: de-extinction is unethical because it involves an arrogant disrespect for the non-human world.
- *The unnaturalness argument*: de-extinction is unethical because it is unnatural.
- *The animal welfare argument*: de-extinction is unethical because it involves inflicting unnecessary pain and suffering on animals.
- *The teachable moment argument*: de-extinction is unethical because there is a valuable lesson to be learned from the extinctions we have caused.

These are all arguments that have been defended in print at one time or another.

To anticipate, our conclusions will be that: (1) the biodiversity restoration argument is by far the most forceful of the different arguments *for* de-extinction; (2) the resource allocation argument is the most forceful argument *against* de-extinction, but has little force against deferred de-extinctions; and (3) all the other above-listed arguments are of limited force or scope.

4.2   THE BIODIVERSITY RESTORATION ARGUMENT FOR DE-EXTINCTION

The argument. The goal of wildlife conservation is to *promote biodiversity.* This means both *protecting extant biodiversity* and *restoring lost biodiversity when and where this is possible* (e.g., as when a local extinction can be undone by translocating extant organisms from one place to another). Given that conservation has this goal, how can de-extinction be justified as a conservationist tool? Very easily. It requires no special justification whatsoever. It is justified, just like other types of conservationist intervention, by its being a means to the end of promoting biodiversity and, more particularly, *of restoring lost biodiversity.* After all, every different species makes its own unique contribution to the Earth's biodiversity, a contribution that is subtracted away from the richness and wonder of the world if the species goes extinct, but that would be added back to it again if the species was to be authentically resurrected and re-established in the wild. De-extinction needs no more justification than this. The same considerations that push us to preserve biodiversity should push us to restore it when it is lost, and de-extinction is nothing but a method of restoring lost biodiversity. De-extinction is, to this extent, continuous with ordinary 'business as usual' for conservationists.

Commentary. Versions of this argument have been endorsed by Crist (2008, p. 290), Cohen (2014, pp. 166–169), Brand (2014), Campbell (2016, pp. 753–754) and Iacona et al. (2017, p. 1041). It rests on two assumptions. The first assumption is that de-extinction technology is genuinely capable of restoring lost biodiversity by bringing back extinct species. Anti-authenticists will, of course, reject this assumption and claim that de-extinction can, at best, only create new, artificial species that *closely resemble* extinct species in outward respects but which are in fact shorn of most of what made the originals valuable (Sandler, 2013, p. 356; Seddon, 2017). If this was correct then de-extinction would merely provide us with a way of replacing lost natural biodiversity with a synthetic simulacrum thereof, not with a way of genuinely recovering what was lost. However, the previous chapter examined the anti-authenticist position in detail and concluded that it is implausible. We will say no more about anti-authenticism here.

The second assumption is that the conservationist's goal of restoring lost biodiversity applies *unrestrictedly*, not just to species that have become locally extinct in some part of their former range, but also to species that have been completely extirpated. This assumption is easy to defend. As a species makes its sorry way down the IUCN Red List, its status changing from *Vulnerable* to *Endangered* to *Critically Endangered* to *Extinct in the Wild*, the task of saving it and of restoring it to a thriving state becomes ever more technically challenging. Usually, however, this is not a reason to give up—not while there is still realistic hope of success. The advent of de-extinction technology means that there can still be hope, even when the species' status shifts to *Extinct*. It would make no sense to arbitrarily stipulate that a species must be abandoned as a lost cause the moment it becomes, say, *Endangered*. It makes no more sense to stipulate that it must be abandoned the moment it becomes *Extinct*. If it is worth fighting to save a species threatened with extinction, then, all else being equal, it is also worth fighting to bring it back from extinction, provided this can be done. If in losing a species we would lose something of great value—a thing that contributes uniquely to the richness of the world, a biological jewel without which Earth would be less wonderful than it was before—then we have reason to fight to save it. By the same token, if we lose that fight and the species goes extinct, then we have a reason to try to resurrect it if we can, circumstances and resources permitting.

The key thing to notice about the biodiversity restoration argument is that it rests on an assumption that conservationists are already wedded to—namely, the assumption that biodiversity is valuable and worth restoring if it is lost. This puts the conservationist who would deny that there can be a lot of value to be gained from resurrecting a species in a very difficult dialectical position. It is for this reason that the biodiversity argument is, in our opinion, an especially forceful argument.

Of course, there are limits to what the biodiversity argument proves. It doesn't prove, all by itself, that conservationists must embrace de-extinction. After all, for all it shows the benefit to be gained from de-extinction—namely, the recovery of lost biodiversity—is not worth the associated costs—including resource costs, animal welfare costs, moral hazard costs, and the costs that would come with our 'disrespecting' the non-human world. Therefore, before we can form a final judgement about whether conservationists should embrace de-extinction, we first need to survey all the other arguments on both sides of the debate.

4.3   THE JUSTICE ARGUMENT FOR DE-EXTINCTION

The argument. The human species has, by causing the extinctions of other species, done great wrong against these species (or against the members thereof). We have a moral duty to right these wrongs for which we were responsible; to make amends; to undo the harm we have done—and this entails that we reverse anthropogenic extinctions when and where we can. In short, we owe a debt of justice to the species we have exterminated, and we are morally obliged to resurrect them for this reason.

Commentary. This is perhaps the most popular of all the different pro-de-extinction arguments (Archer, 2013; Brand, 2013; Cohen, 2014; Jebari, 2016). However, in our opinion it is a lot less logically forceful than the biodiversity restoration argument. This is because it rests on tendentious assumptions that conservationists are not wedded to and that they could easily opt to reject.

To see this, consider the extinction of the passenger pigeon. Who were the perpetrators of this extinction? They were an army of people with guns, traps and nets who slaughtered the birds by the million in the American northeast in the late 1800s. Who were (and are) the victims of this extinction? The answer, surely, includes *people alive today*, along with all people of the future. After all, the extinction of the passenger pigeon has deprived us of the opportunity to see and experience one of the greatest, most thrilling natural wonders of the world. Why should we resurrect the passenger pigeon? Well, at least partly so as to restore this awe-inspiring component of the Earth's biodiversity, and enable our descendants to know what it is like to see a flock of three billion birds blotting out the sun for hours at a time.

This is a reason to resurrect the species—perhaps even a good reason—but, importantly, it is a reason that comes, not from the justice argument, but from the biodiversity restoration argument. The justice argument requires us to be cast in the role, not of victim, but of villain. It relies on the assumption that we are, if not exactly the guilty party in the extinction of the passenger pigeon, then at least a proxy for the guilty party where issues of justice are concerned (Cottrell, Jensen, & Peck, 2014, p. 8). It assumes that a moral debt to the passenger pigeon species was incurred by those who shot and netted and clubbed the birds, and that this debt has been passed forward, down the generations, to people of the current day. In other words, it relies on the assumption that the child inherits responsibility for the ecological crimes of the parent. The implication is that a human babe is born into the world with the culpability from all past anthropogenic extinctions hanging albatross-like around her little neck—the environmentalists' version of the Christian doctrine of original sin. We are guilty by association, because—however much we abhor what was done to the passenger pigeons—we are tied by bonds of species membership to those who actually did the deed.

We do not want to claim here that this assumption is necessarily false. Maybe it is true. Maybe when we destroy a species, we don't just harm our descendants once, by leaving them with a less wonderful world to live in, but twice, by also burdening them with the moral duty to undo what we have done. Here we merely wish to point out that it is far from obvious that this assumption is true, and nor is there any reason why a conservationist must accept that it is true. (In fact, we think conservationists would be well advised *not* to accept it.) A conservationist can be adamant that biodiversity is valuable, and be motivated by this belief to promote biodiversity, while not for one moment buying into the idea that moral duties to resurrect extinct species have been passed down to us from our extinction-causing forebears.

The justice argument also rests on a second tendentious assumption—namely, that there is some non-human victim of a species' extinction to which a moral duty is owed. Again, consider the passenger pigeon. The extinction of the passenger pigeon species was brought about by the killings of untold millions of individual birds. Unfortunately, we can do nothing at all to bring these individual birds back or make amends to them. They are gone for good. At most we can create new members of their species. But if one kills, say, a duck, then one does not make amends to the dead duck by using synthetic biology to create a new duck. What good would that be to the duck who died? By the same token, what good is a newly created, de-extinct passenger pigeon to a passenger pigeon who was killed by a hunter a century and a half ago? Perhaps it would be of some small comfort to a human being who was about to be murdered to know that his death would be memorialized, many years later, by the creation of a new human being. But it is grossly anthropomorphizing passenger pigeons to project such sentiments into their pigeon brains.

Since the harm that was done to individual passenger pigeons when they were killed cannot be undone, there cannot be a duty to undo these particular harms. ('Ought' implies 'can', and so there cannot be a duty to do the impossible.) But if a moral debt is owed then it must be owed to someone or something—so to whom or to what is it owed? It might be said that the victim of the extinction to which we owe a moral debt is *the passenger pigeon species as whole*, rather than the slaughtered birds considered individually. However, a species is a taxonomic group of organisms, and such taxonomic groups are not sentient even if the organisms in them are sentient. A species itself has no desires, or goals, or interests, or intentional states (Kasperbauer, 2017; Sandler, 2013, p. 355). It neither wants to remain extant nor wants to go extinct, for its wants are a nullity. It no more has wants, or interests, than does a stone or the number six.[1]

In short, if the human race incurred a moral debt when the passenger pigeon was exterminated, then it is unclear to whom or to what this debt is owed. Could there be a solution to this mystery? Maybe.[2] Here our point is only that there is ample room for a conservationist to be sceptical. If there is no one and nothing to whom the debt is owed, then the debt does not exist, and so the justice argument rests on a false assumption.

The biodiversity restoration argument for de-extinction is, we think, far superior to the justice argument because it doesn't require these problematic assumptions to be made or defended. It doesn't rely on our being able identify a victim of an extinction to whom a moral debt is owed. Nor does it assume that the duty of paying this debt has somehow been inherited by people of the modern day even in cases where we had nothing to do with the extinction and would have stopped it if we could.[3] We think proponents of the justice argument needlessly weaken and cloud their position by burdening themselves with these assumptions. When our forebears extirpated species like the passenger pigeon they (knowingly or not) harmed people of all future generations, including ourselves, by depriving us of the opportunity to experience the species in question, and by leaving the world a less gloriously biodiverse place than it used to be. If the extinctions had victims, then these victims include *us*. By reversing these extinctions we will be undoing these harms that were done to ourselves and our descendants. For a conservationist, that's motivation enough for resurrecting extinct species. Tendentious claims about inherited moral duties are surplus to requirements and need not be brought into it.

4.4   THE ECOLOGICAL RESTORATION ARGUMENT
FOR DE-EXTINCTION

The argument. Extinct species often provided important ecological services—such as ecosystem engineering, predation, frugivory, herbivory, seed dispersal, and pollination—that would greatly benefit extant species (including human beings) if they could be restored. Conservationists should support the resurrection of such keystone species as a way of restoring these services (McCauley, Hardesty-Moore, Halpern, & Young, 2017).

Commentary. This is an argument of very limited scope, for several reasons. First, it only applies to extinct species *that would in fact provide valuable ecological services* if they were resurrected. It is not applicable in all the numerous cases where the extinction of a species hasn't caused any untoward ecological effects.

Second, there will usually be much easier and more cost-effective ways of restoring lost ecological services than by resurrecting the species that originally provided them. Often it will be possible to achieve the desired ecological effect by using some other, extant species to provide the service in question—as, for example, by using giant tortoises to provide the browsing services once provided by now-extinct Hawaiian species of giant, flightless ducks and geese (Burney, Juvik, Burney, & Diagne, 2012), or by using Asian elephants as ecological proxies for extinct North American mammoths (Donlan et al., 2006, pp. 669–670). The techniques of synthetic biology will be called for only in cases where no such extant ecological proxy exists. Moreover, even in such cases it might be possible to achieve the desired result by making relatively small genetic modifications to an extant species instead of by resurrecting the extinct species. For example, we needn't recreate authentic woolly mammoths in order to restore the ecological services they provided in the subarctic, since, as we saw in Chap. 2, it might suffice to re-engineer Asian elephants for cold-tolerance by giving them just a few key woolly mammoth genes.

Third, a species that provided valuable services before it went extinct will not necessarily provide the same services if it is resurrected because of environmental and ecological changes that have occurred in the interim, and because of the profound, and often unpredictable, impacts such changes might have on a species' ecological role (Caro, 2010, pp. 138–142, 151–153; Cohen, 2014, p. 169).

The assumption that long-lost species can restore historic functions rests on a static view of gene–organism–environment relationships. It fails to appreciate the many changes and adjustments that take place after a species has disappeared. After the loss of a species, there is often a cascade of functional changes that can be very difficult if not impossible to reverse. (Kohl, 2017, p. 16)

For these three reasons the ecological restoration argument will seldom provide a strong justification for resurrecting an extinct keystone species. Although it might be forceful in rare cases, it is an argument of much more limited scope than the biodiversity restoration argument.

4.5   THE SMART POLITICS ARGUMENT FOR DE-EXTINCTION

The argument. In so far as the goal of conservation biology is merely to slow the rate of biodiversity loss, it is a remorselessly depressing and dispiriting discipline. Its ethos is one of doom and gloom, of loss and blame, of despair and guilt, of never-ending crises. This negativity, justified though it may be, is politically debilitating. It inculcates crisis fatigue, a sense of hopelessness, and 'learned helplessness' (Redford & Sanjayan, 2003, p. 1473; Swart, 2015, p. 67). Human psychology being what it is, it causes people to turn their attention to other, more cheerful topics. In order to achieve buy-in from the public and attract much needed political support, conservation biologists need to pursue their cause "from a position of inspiration, not blame", and "offer humans the means to envision a positive and achievable vision of the future" in order to "gain the hearts, mind and—most importantly—the actions of the world's people" (Redford & Sanjayan, 2003, p. 1474). De-extinction can be a vital part of the required paradigm-shift. Indeed, it is perfectly suited to the purpose. The resurrection of a charismatic lost species and its re-establishment in the wild is precisely the type of positive, up-beat, exciting event that will grab the attention of the media and grip the imaginations of lay people (Donlan, 2014, p. 27). Thus, de-extinction is *smart politics.* It is an "umbrella technology" (Crist, 2008, p. 289), that will benefit innumerable extant species by dispelling the all-pervasive gloom that hangs over wildlife conservation and by helping to get the public firmly behind conservation projects. It will shift the conservation story "from negative to positive, from constant whining and guilt-tripping to high fives and new excitement" (Brand, 2014).

Commentary. The smart politics argument has limited force and scope for several reasons.

Firstly, and most importantly, it is not obvious that de-extinction would, on balance, be 'smart politics'. Would some members of the public become enthused about conservation as a result of the high-profile de-extinction of a charismatic species? Yes. But there will, inevitably, also be other people who, rightly or wrongly, see de-extinction in a negative light, and who will therefore be inclined to view conservationism in a negative light too if conservationists endorse and embrace de-extinction. Will the net public relations effect of de-extinction be positive or negative? This is an empirical question, and we don't pretend to know the answer, but the point is that there is no *a priori* reason to think the net effect will be strongly positive.

The second point is closely related to the first. A side effect of a high-profile de-extinction project would be to undermine the public's acceptance of the idea that 'extinction is forever'. This could help foster complacency about the current biodiversity crisis, and thereby seriously undermine conservationism's political support. We will return to this point below, when we discuss the 'moral hazard argument' The moral hazard argument and smart politics argument are antagonistic arguments that support diametrically opposing conclusions, with the moral hazard argument suggesting, in effect, that de-extinction is 'dumb politics', not 'smart politics'. To the extent that each of these arguments is forceful it cancels out the force of the other.

Thirdly, there is a respect in which the smart politics argument cheapens de-extinction, by making it out to be a public relations gimmick peripheral to conservationism's core business of protecting and restoring biodiversity. The biodiversity restoration argument doesn't do this. It instead justifies de-extinction on the basis of its ability to contribute directly to this core business, by bringing back lost biodiversity. The biodiversity restoration argument would, for this reason, still be applicable even if there was already robust public support for de-extinction and no need for public relations exercises. Of course, this is not to suggest that public relations exercises are unimportant. Perhaps de-extinction would be invaluable in building a strong groundswell of support for conservationism, just as proponents of the smart politics argument suggest. But it is to suggest that the smart politics argument provides a much more tenuous and unreliable justification for de-extinction than does the biodiversity restoration argument. The smart politics argument provides a justification for de-extinction that is contingent on public attitudes of the day towards conservationism and biotechnology. It seems to imply that the best candidates for de-extinction will usually be the cutest or the most symbolically important, not the ones that made an outsized contribution to the Earth's biodiversity. It also implies that de-extinction will probably be subject to a strong law of diminishing returns, since the first high-profile de-extinction will presumably engender a lot more of a transformative, feel-good glow than, say, the nineteenth.

4.6   THE ARGUMENT FROM COOLNESS
FOR DE-EXTINCTION

The argument. It would be cool to resurrect lost species like the woolly mammoth—and so we should do it.

Commentary. The argument from coolness is sometimes mentioned as providing a rationale for de-extinction (Sherkow & Greely, 2013, p. 33), and where the actual driving motivations behind current de-extinction projects are concerned, perceptions of 'coolness' are surely of considerable significance. However, from a logical perspective the argument from coolness appears to be reducible to some of the other pro-de-extinction arguments we have already mentioned. Two versions of the argument from coolness can be distinguished. The first version is premised on the idea that extinct species like the woolly mammoth were cool (or wonderful, or awe inspiring) when they trod the Earth long ago, so that we would be recovering something cool by resurrecting them. This version of the argument from coolness is nothing but a version of the biodiversity restoration argument in which one particular aspect of a lost species' aesthetic value to humanity, namely, its coolness, is emphasized as being a reason why it was bad to lose it and why it would be good to bring it back again. The second version of the argument from coolness focuses not so much on the coolness of the resurrected species itself as on the (putative) coolness of *the act of resurrecting a species*. Whether resurrecting a species is cool or not is surely a highly subjective matter. (Many will deem it creepy, not cool.) But if it were true that a substantial fraction of the public thought de-extinction was cool then this would give us a version of the smart politics argument, with the idea being that de-extinction projects are of conservation value because they offer a way of getting the public excited about conservation.

4.7   THE BALANCE RESTORATION ARGUMENT
FOR DE-EXTINCTION

The argument. We need to resurrect extinct species in order "to restore the balance of nature that we have upset" (Archer, 2013).

Commentary. The idea that there is any such thing as 'the balance of nature', or a long-term equilibrium to which natural systems tend, is highly suspect and now almost universally repudiated by ecologists (Botkin, 1992; Pickett, 2013; Simberloff, 2014; Wu & Loucks, 1995). Natural systems are chaotic on almost every level and subject to perpetual change. Because the balance restoration argument rests on an extremely tendentious premise it is a very weak argument for de-extinction.

This completes our survey of common arguments *for* de-extinction. We turn now to arguments *against* de-extinction.

4.8   THE RESOURCE ALLOCATION ARGUMENT
AGAINST DE-EXTINCTION

The Argument. Conservation is severely underfunded (McCarthy et al., 2012; Waldron et al., 2013). Scarce conservation resources need to go where they will deliver the biggest bang for the buck—namely to habitat protection and to preventing species going extinct in the first place. De-extinction projects won't deliver an appreciable bang for the buck any time soon. They carry prohibitive upfront costs (Callaway, 2016), and would, even if successful, also burden conservationists with the long-term costs of monitoring and protecting resurrected populations (Bennett et al., 2017; Cottrell et al., 2014, p. 9). Hence de-extinction is a wasteful misallocation of resources (Bennett et al., 2017; Blockstein, 2017; Ehrlich & Ehrlich, 2014; Iacona et al., 2017; Sandler, 2017). In the words of the editors of the *Scientific American* (2013):

> A costly and flamboyant project to resuscitate extinct flora and fauna in the name of conservation looks irresponsible: Should we resurrect the mammoth only to let elephants go under? Of course not.

Commentary. In evaluating this argument it is important to carefully distinguish among the following different types of de-extinction projects:

1. *Deferred de-extinction projects.* Such a project would have three phases. In phase 1, any surviving tissue samples from an extinct species (including samples of bone, teeth, feather and skin stored in museums) would be cryogenically preserved, to stop DNA degradation (which is inexorable at room temperature). These samples should be as genetically diverse as possible. Where possible, they would ideally be collected from an endangered species *well prior* to its extinction, as a precautionary measure (Benirschke, 1984; Crist, 2008; Ryder et al., 2000). Ideally, they will include gametes (ova and sperm), as well as somatic cells, to open up additional de-extinction pathways. In the case of plants, they should obviously include *seeds*. Samples of gut flora and other microbial symbionts should also be frozen where possible (McCauley et al., 2017, p. 1006). Phase 1 will ideally also include gathering and caching as much useful information as possible about the biology and ecology of the species. Phase 2 is a 'waiting' phase. It may continue for decades or even for centuries, during which time de-extinction technology is likely to become radically more powerful and exponentially cheaper and easier to use. During this phase various inexpensive preliminary measures, such as genome sequencing, might be taken by way of preparing the ground for the species' eventual de-extinction. Phase 3 is the resurrection phase. It is to begin only if and when money and resources for resurrecting and conserving the species can be spared from other more pressing conservation needs.

2. *Self-funding projects.* These are projects that pay their own way, by getting their money from sources that wouldn't otherwise fund conservation work. Possible non-traditional sources of funding include private or corporate sponsors with a specific interest in one particular extinct species (Brown, 2016; Sandler, 2017, p. 2; Swart, 2015, p. 65), zoos that would profit from displaying de-extinct animals (Cottrell et al., 2014, p. 8), a market for de-extinct pets (Carlin, Wurman, & Zakim, 2013, p. 5; Welchman, 2017), or tourism (Whittle, Stewart, & Fisher, 2015).[4] The resource allocation argument assumes that conservation funding is a 'zero sum game', and this assumption will be invalidated to the extent that de-extinction can attract funding from non-traditional sources.

3. *Umbrella projects.* These are projects wherein de-extinction is part of a larger strategy with the primary goal of protecting and restoring *extant* biodiversity. For example, the de-extinction of a keystone

species might benefit numerous extant species by restoring a lost ecology. This was the idea behind the ecological restoration argument, above. Or the de-extinction of a charismatic, flagship species, and all the hype and media attention it attracts, might benefit extant species by "garnering the support and interest of average citizens" (Donlan, 2014, p. 27), by engendering public support for habitat protection, by creating enthusiasm for environmentalist causes, and by attracting new skills and talent into conservation (Cohen, 2014; Crist, 2008, p. 289; Iacona et al., 2017; Jones, 2014). This is the idea behind the smart politics argument. Or a de-extinction project might be a technological 'moon-shot' that benefits many extant species by creating new conservation technologies of broad applicability (such as the 'artificial womb' technology mentioned in Chap. 2).

4. *Leech projects.* These are de-extinction projects that siphon resources away from other conservation work and which are counterproductive to conservation's overall goal of preserving and restoring biodiversity.

Of these four types of de-extinction projects, the argument from misallocated resources undermines only leech projects. However, of the remaining three types of projects it is plausible that both self-funding projects and umbrella projects will be uncommon. They are likely to be possible only if the species in question is especially charismatic or especially ecologically important. It is hard to imagine more than a dozen or two such projects ever going ahead—a drop in the ocean where the current biodiversity crisis is concerned. We conclude on this basis that, in so far as de-extinction projects are to play any very significant role in restoring lost biodiversity, the projects in question will by and large *be deferred de-extinctions.*

This is a point that is seldom remarked upon, but which we believe has profound implications for how conservationists should think about de-extinction. Discussions of the ethics of de-extinction typically rest on two misconceptions. The first is that if de-extinction is something that should be done *at all*, then it should be done *now*. The second is that the main targets of de-extinction will be species, like the woolly mammoth, the aurochs, or the passenger pigeon, that are already extinct as of the present day. We think both these ideas are wrong-headed. There are compelling reasons not to engage in de-extinction now on anything more than a very small, experimental scale. First, de-extinction technology is still in its earliest infancy, and hence prohibitively costly and difficult to use. Second, given exponential rates of improvement in the technology, costs will drop precipitously if we simply *wait*. Third, we can afford to wait. Where de-extinction is concerned there are no pressing deadlines. Biological samples from an extinct species can potentially be kept 'on ice' for decades or centuries before being used to bring a species back, and so conservationists have the luxury of time.[5] Fourth, conservationists have no such luxury where extant species are concerned—not with more than five thousand critically endangered species currently teetering on the brink of extinction (IUCN, 2017). Hence it is extant species, not extinct species, that need to be the absolute focus of our conservation efforts for the immediately foreseeable future.

These are all good reasons for thinking we shouldn't be spending precious conservation resources on resurrecting extinct species *now*. However, they are not good reasons for thinking de-extinction shouldn't be an important part of long-term conservation planning. Conservation biology involves taking a long view, over the sorts of hundred-year timescales required for forests to grow, ecosystems to stabilize, and slow-breeding endangered species to recover their numbers. De-extinction's potential contribution to wildlife conservation needs to be assessed with these sorts of timescales in mind—timescales which mean the present-day costs and limitations of the technology are unimportant. For conservationists of today, de-extinction should be all about using that little-noticed and under-appreciated piece of conservation hardware, *the freezer*. It should be about cryogenic preservation; about stocking a frozen ark. Where de-extinction is concerned, the conservationist's motto should be 'freeze now and resurrect later'. When conceived of in this way, de-extinction is not about trying to use our existing, immature biotechnology to bring back extinct species *today*. Rather, it is about providing synthetic biologists of the future—whose biotechnology will be astronomically more powerful than ours—with the biological samples and information they will need to undo extinctions we have been unable to prevent. It is about taking easy and inexpensive measures to give them options they would otherwise lack, thereby empowering them to correct our mistakes as they see fit (Crist, 2008, pp. 289–290). In Eileen Crist's words, "how could the banking of cell-lines from the world's frogs be regarded as anything other than a rational safeguard, best undertaken immediately?" (2008, p. 289).

As so understood, de-extinction is not a 'backwards looking' conservation technique. It is true that it might be used to undo some mistakes of the past, by allowing us to bring back a few of the species—like the aurochs, or the passenger pigeon—we have already wiped out. But its most important application will be to extinctions that haven't happened yet. Because of inadequate resourcing, conservationists are often forced to triage endangered species, dividing them into those that should be saved and those that must be let go. Such wrenching dilemmas are set to become much more common over the next few decades as the Holocene mass extinction event gathers steam and as climate change and ocean acidification tighten their grip on the planet. De-extinction technology provides conservationists with an additional, 'freeze now and resurrect later' option in such cases. Although this option will always be much worse than saving a species from extinction, it is still far preferable to letting the species disappear without paving the way for conservationists of the future to bring it back.

Not only does de-extinction technology provide conservationists with a new emergency fall-back option when triaging is necessary, it also has important implications with respect to management decisions about *which particular species* should be saved from extinction and *which techniques* should be used to save them. For example, suppose we have a choice between saving species X or species Y from extinction, but where X is much more amenable to de-extinction than Y (perhaps because Y poses trenchantly difficult animal husbandry challenges not posed by X). In this case we should, all else being equal, save Y from extinction and let X undergo a 'managed extinction' (with the hope being that X's extinction will only be temporary). Or suppose we have a choice between using either method M or method N to save a species, S, from extinction, where M is considerably less expensive, but riskier, than method N, and where S is a species that would be readily amenable to de-extinction if it were to go extinct. Before the advent of de-extinction technology we would have had very strong reasons to err on the side of caution, by using N. The advent of de-extinction technology makes the accidental loss of S less catastrophic, by opening up the possibility of its being resurrected. (It buys us insurance, so to speak.) The technology therefore shifts the cost-benefit calculus more in favour of using the less expensive method, M, enabling the money saved to be diverted to other conservation projects (Iacona et al., 2017, pp. 1043–1044).

4.9   THE REWILDING ARGUMENT
AGAINST DE-EXTINCTION

The argument. In Chap. 1 we distinguished 'nominal de-extinctions'—which resurrect a species just barely and perhaps only fleetingly—from 'Least Concern de-extinctions'—which establish thriving new populations of a species in the wild. Achieving a nominal de-extinction would be an impressive feat of synthetic biology, but that's all. Achieving a Least Concern de-extinction would also require overcoming a series of major problems of animal husbandry, conservation biology and population genetics. Technical hurdles to be negotiated would include:

1. Almost any conceivable Least Concern de-extinction project would need to include a captive breeding phase (assuming the species to be resurrected was an animal, rather than a plant). However, many animals are not amenable to captive breeding (for example, migratory sea birds), and those that can be captive-bred often become genetically adapted to captivity, preventing their successful reintroduction into the wild (Snyder et al., 1996).

2. Creating a genetically healthy breeding population would entail incorporating genetic material from a genetically diverse sample of the pre-extinction population. However, this will often be impossible because of a paucity of DNA sources or because the pre-extinction population was itself genetically depauperate (Steeves et al., 2017, p. 1037).

3. In order for a de-extinct population to be invulnerable to gradual erosion of its genetic diversity by genetic drift (which would result in inbreeding depression and a reduced capacity to adapt to environmental change) it must be *large*. Indeed, the requisite population size will typically be upwards of 5000 organisms, an extremely onerous target in most cases (Steeves et al., 2017, pp. 1034–1035).

4. If the resurrected species is an animal that transmits behavioural information from generation to generation via learning (e.g., information about foraging methods, predator avoidance or migration routes) then the animals of the first generation will have no conspecific role-models to learn from, greatly diminishing their survival prospects. (Similar issues arise in relation to imprinting.)

5. The de-extinct organisms must be released into a wild habitat where they can thrive, but for many species no such habitat exists because

habitat-loss drove the species extinct in the first place, or because the drivers of extinction (e.g., disease, or invasive predators, or human interference) are still operative in their former habitat, or because another species now occupies the extinct species' niche, or because environmental and ecological changes that have occurred during the species' absence mean the species is now poorly adapted to current conditions (Kohl, 2017, p. 16; Robert et al., 2017).

6. The habitat the de-extinct species is released into must not itself be in danger from climate change, sea-level rise, ocean acidification, pollution, or other looming anthropogenic threats.

7. It must be possible to reverse the translocation of the de-extinct organisms into the wild if something goes wrong—e.g., if the species turns out to be invasive in the current-day ecological context, or a vector for disease transmission (IUCN/SSC, 2016, p. 144). For some species (e.g., birds that are highly mobile) it might be very difficult to reliably 'undo' a wild release.

8. The rewilding of the species must be politically feasible in light of public opinion and the interests of powerful lobby groups.

According to what we call the 'rewilding argument' against de-extinction, factors like (1)–(8) render de-extinction bankrupt as a conservationist tool (Ehrlich & Ehrlich, 2014; Pimm, 2013; Scientific American Editors, 2013). The thought is that at best synthetic biology can only elevate the status of a species from 'extinct' to 'extinct in the wild', still leaving all the hard problems of animal husbandry and translocation to be solved by already badly overstretched conservation biologists. The IUCN Red List presently records 68 species and subspecies as being 'extinct in the wild' (IUCN, 2017). The last thing resource-strapped conservation facilities and conservation workers need right now is the burden of more such species to care for.

Commentary. There are several weaknesses in this argument.

Firstly, and most importantly, there is no denying that conservationists are presently overburdened with species that need captive rearing and rewilding, but the upshot is not that de-extinction is bankrupt as a conservationist tool. Rather, it is that a 'freeze now and resurrect later' de-extinction policy should be pursued. With rare exceptions, deferred de-extinctions rather than precipitate de-extinctions will be the order of the day given that we in the midst of an ongoing biodiversity crisis.

Second, it is overstating things to say that achieving a Least Concern de-extinction will *always* be problematic because of factors like (1)–(8).

Some species are relatively easy to captive breed and rewild, and others, nigh on impossible. Factors like (1)–(8) should be crucial considerations in deciding *which* species to resurrect, but they don't leave synthetic biologists without any good candidates to choose from. For example, the aurochs is a relatively unproblematic candidate for de-extinction, since none of factors (1)–(8) appear to pose insuperable difficulties where aurochs are concerned. In contrast, the passenger pigeon is at least somewhat problematic as a de-extinction candidate because, even though Ben Novak's passenger pigeon de-extinction plan mitigates most of factors (1)–(8), big questions remain (for example, the rewilding of the passenger pigeon might not be gladly received by the horticultural sector).

Thirdly, there are measures synthetic biologists can take in order to ameliorate some of factors (1)–(8). Recall from Chap. 1 that the reproductive fitness of a de-extinct population could be enhanced by performing a 'discriminating de-extinction', in which deleterious genes would be deliberately omitted from the reconstructed genepool. The omission of deleterious genes would dramatically reduce the population's susceptibility to inbreeding depression, helping to ameliorate factor (3). Furthermore, de-extinct organisms produced in this way would be genetically superior to the members of the pre-extinction population in so far as they would have an unnaturally low mutation load. The resulting reproductive advantage would gradually wear off over subsequent generations, but in the meantime it would help to counterbalance reproductive disadvantages stemming, say, from the de-extinct organisms not having mature conspecifics to learn behaviour from, or from their being adapted to environmental conditions that no longer obtain. Also, some of the problems that afflict ordinary captive breeding programs would be less serious in the context of a de-extinction project. For example, erosion of genetic diversity, and genetic adaptation to captivity, both of which are normally extremely serious problems, could be readily countered within a de-extinction project by occasionally topping up the genepool of the captive reared population with new organisms having whatever desirable genes have been lost.

4.10   The Moral Hazard Argument Against De-extinction

The argument. The received wisdom that 'extinction is forever' has long provided conservationists with a politically effective 'red line' they could use to oppose practices that risked causing extinctions. De-extinction will dramatically weaken conservationists' hand by refuting this piece of received wisdom, and blurring or erasing the red line. It will create a 'moral hazard' by enabling the extinction of a species to be excused on the basis that the harm is reparable at a later date. It will de-incentivize the public from caring about the underlying causes of extinctions, such as habitat fragmentation, and will offer "unscrupulous developers a veil to hide their rapaciousness, with promises to fix things later" (Pimm, 2013). This is among the most common objections to de-extinction (Ehrlich & Ehrlich, 2014; Iacona et al., 2017; Kohl, 2017; Pimm, 2013; Sherkow & Greely, 2013).

Commentary. There are several problems with the moral hazard argument.

First, moral hazards are ubiquitous (Brand, 2014; Donlan, 2014). In general, any new method for alleviating a harm creates a moral hazard by de-incentivizing precautions against causing that harm. But, obviously, it doesn't follow that all new methods for alleviating harms are ethically pro-scribed. To the contrary, they won't be proscribed so long as their benefits outweigh the moral hazard they create. That de-extinction carries a moral hazard is obvious, but that this moral hazard outweighs de-extinction's conservation benefits (first and foremost, the restoration of lost biodiver-sity) is not obvious.

Second, the moral hazard argument against de-extinction has little force against deferred de-extinction projects, because they pose little moral hazard. The cryobanking of cell-lines from endangered species has been going on since the establishment in the 1970s of San Diego's 'Frozen Zoo'. The Frozen Zoo was opposed at its inception on the basis of the moral hazard it would allegedly create, but this moral hazard didn't mate-rialize (Brand, 2014). Low-profile projects to cryobank DNA samples and cell-line haven't undermined public acceptance of the idea that extinction is a red line that shouldn't be crossed.

Third, as was pointed out above, the moral hazard argument collides head-on with the smart politics argument, with the former saying that de-extinction's political effects will be bad, and the latter, that they will be good. It is unclear (to us) which argument is the most forceful, but it is at least clear that de-extinction's political consequences will be complex, and that some (if not all) of its bad political effects will be counterbalanced by good political effects.

Fourth, the reality appears to be that de-extinction is inevitable, as many commentators have pointed out (Jones, 2014, p. 23; O'Connor,

2015, p. 191; Seddon, 2017, p. 994; Zorich, 2010). Multiple de-extinction projects are already well advanced, as we saw in Chap. 2, and there appears to be no realistic prospect whatsoever of a global moratorium on de-extinction at this point. This being so, the moral hazard of de-extinction is unavoidable. This doesn't mean the moral hazard argument is unsound: for obviously it might be that de-extinction *shouldn't* go ahead because of the moral hazard it will create, even if it in fact *will* go ahead. But it does entail that the moral-hazard argument might have a limited shelf-life. Once a few high-profile de-extinctions have taken place and the public's belief that extinction is forever has been well and truly shattered, subsequent de-extinctions won't carry much of a moral hazard (the damage having already been done), and so the moral hazard argument will lose its force. In other words, the moral hazard argument will have much more force against the first de-extinction project than against, say, the ninety first.

4.11   THE HUBRIS ARGUMENT AGAINST DE-EXTINCTION

The argument. The case for de-extinction rests on the hubristic assumption that the consequences of our tampering with genes and with life can be safely predicted and controlled. This assumption is false. Biological systems are inherently unpredictable and prone to non-linear, cascading changes (Nes & Scheffer, 2004). The best-laid plans of synthetic biologists, therefore, won't survive contact with reality. As in *Jurassic Park*, something will go wrong. For example, we might, in the act of resurrecting some species, unwittingly also resurrect a dangerous retrovirus hiding in its genome (Sherkow & Greely, 2013). Chimeric test-organisms not intended for release might escape to wreak environmental havoc. Or de-extinct organisms might turn out to be invasive 'pests from the past' in modern-day ecosystems (Kasperbauer, 2017; Minteer, 2014). In short, de-extinction is *inherently risky*. The Precautionary Principle dictates that such risky actions not be taken.

Reply. Such risks are real but not unique to de-extinction (Donlan, 2014, pp. 25–26). They are not deal-breakers. For conservation biologists and genetic engineers, they are everyday facts of life. Conservation biologists commonly translocate species back into habitats from which they have long been absent, and there are well-developed guidelines for mitigating the risks (e.g., of invasiveness, disease transmission, and unexpected species interactions) attending such translocations (IUCN/SSC, 2013;

Seddon, Moehrenschlager, & Ewen, 2014). A de-extinction project that puts a resurrected species back in the wild is nothing but a type of translocation project (Seddon, 2017; Seddon et al., 2014), and the "well-established standards for species reintroduction projects provide a solid foundation on which de-extinction can be built" (Jørgensen, 2013, p. 719). There are also extensive guidelines for containing genetically modified organisms, and for carefully managing their staged release. A de-extinction project will not be hubristic if it adheres to protocols developed by experts with a clear-eyed understanding of the risks and uncertainties, and who, in drawing up the protocols, err on the side of caution. Such protocols have already been developed (IUCN/SSC, 2016). Projects not adhering to these protocols should indeed be opposed by conservationists.

4.12   THE PLAYING GOD ARGUMENT AGAINST DE-EXTINCTION

The Argument. The root cause of most of the environmental damage our species has wrought is our willingness to 'play god' by bending nature to our will with technology. An environmental ethic embodying a proper respect for the biosphere and a proper understanding of humanity's place in nature will proscribe such technological interventions. It will require us to scale things down; to stop trying to manage and control the non-human world; to tread more lightly on the land. Since de-extinction involves the heavy-handed use of technology to manipulate and control life, it is flatly inconsistent with such an ethic.

Here are two examples of this argument in print:

> At bottom, de-extinction is more experimental and novelty-producing than it is restitutive or restorative; its ideal is less a call for humans to scale things down to make room for other forms of life than it is a summons to keep scaling up our technological and managerial interventions in their worlds. It is this feature of de-extinction that is most inescapable ... and it is for this reason that it is so deeply objectionable. (Diehm, 2017)

> Attempting to revive lost species is in many ways a refusal to accept our moral and technological limits in nature.... Leopold was aware of our tendency to let our gadgets get out in front of our ethics. "Our tools," he

cautioned in the late 1930s, "are better than we are, and grow better faster than we do. They suffice to crack the atom, to command the tides. But they do not suffice for the oldest task in human history: to live on a piece of land without spoiling it." The real challenge is to live more lightly on the land and to address the moral and cultural forces that drive unsustainable and ecologically destructive practices.... It cuts against the progressive aims of science to say it, but there can be wisdom in taking our foot off the gas, in resisting the impulse to further control and manipulate; to fix nature. (Minteer, 2014, p. 261)

Reply. This argument proves too much. It opposes de-extinction by way of opposing technological management of the natural world more generally, but in many cases the only alternative to aggressive, large-scale conservationist interventions is to let unique, hugely precious, million-year-old lifeforms or ecosystems be destroyed by processes we have already (stupidly) set in motion. For example, European settlers of the 1800s foolishly played god by bringing possums, deer, stoats, ferrets, weasels and cats to New Zealand, where they became invasive pests, devastating New Zealand's forests and helping drive numerous bird species to extinction. Many more species would have gone extinct but for the work of conservationists—work that has largely consisted of eradicating invasive species from offshore islands, and, more recently, from parts of the mainland. New Zealand's government has now approved a plan to rid the entire country of rats, stoats and possums by 2050. This will require developing and deploying new technologies on an immense scale, including self-resetting traps, species-specific toxins, and, most promisingly, genetically engineered gene drives designed to skew the sex-ratios of invasive species so that their populations collapse (Hansford, 2016). Only by intervening on such a grand scale can conservationists finally close the lid on the ecological Pandora's box that the Victorian settlers opened more than a century ago.

New Zealand is not a unique case. Invasive species transported from region to region by ships and planes are causing slow-motion environmental catastrophes all around the world. Huge changes to the Earth's climate, the oceans' pH, and global sea levels have already been locked in by our burning of fossil fuels—changes set to unwind to calamitous ecological effect in all corners of the globe over coming decades and centuries. The Earth's human population is now seven billion, and increasing—its appetite for land, insatiable; its hunger for resources, rapacious; its capacity to pollute, prodigious. For all these reasons, and for many more besides (habitat fragmentation, overfishing, etc....), the scene is now set for a Holocene mass extinction event. No 'environmental ethic' worthy of the name would countenance our not doing what we can to save what can be saved from the pending tsunami of extinctions. A fundamental realignment of our species' relationship with the natural world is urgently demanded, just as Minteer indicates in the above quotation. But while we are waiting for this to happen (and no one should hold their breath), intelligently planned conservation interventions will be vital. Techniques such as assisted colonization, intra-species gene de-extinction, facilitated adaptation and interspecies cloning are likely to be indispensable for salvaging important fragments of the Earth's biodiversity from the mass extinction event that is coming (Redford, Adams, & Mace, 2013). So too is species de-extinction (Crist, 2008).

Minteer anticipates this reply to his argument. He makes a major concession, writing that "in many cases intensive and aggressive conservation actions will be required to protect biodiversity in the coming decades" (2015, p. 16). He mentions assisted colonization as an intervention that might be required. But he adds that "Conservation in the Anthropocene must be a balancing act between the pragmatic need for action and the moral wisdom of ecological restraint" (ibid.), with the implication being that de-extinction is for some reason *not* among the aggressive conservation actions that moral wisdom countenances. This begs the question, why not? What is the principled difference between de-extinction on the one hand and assisted colonization on the other, in virtue of which the latter gets the thumbs up from moral wisdom while the former gets the thumbs down? Minteer doesn't elaborate. Perhaps his answer would have to do with the greater risks and unknowns of de-extinction—but then his argument would just be a version of the hubris argument, which has already been dealt with above. Alternatively, his answer might be that de-extinction involves heavy-handed technological management of the natural world, whereas assisted migration doesn't. The problem with this answer is that sometimes heavy-handed technological management is precisely what is needed to save threated biodiversity, or to restore lost biodiversity. The question comes down to what conservationists should do when the only way to save some precious species from being lost to the Earth forever is by resorting to heavy-handed technological intervention. Should they sit on their hands and let the species go, comforting themselves with the thought that they are being wise and not playing god? Or should they intervene, high-minded principles be-damned? Here we can only say that we respectfully disagree with Minteer. We are on the side of intervening. We suspect that most conservationists will be consequentialist enough in their thinking to opt for intervening too. Philosophical principles that would tie our hands in dealing with the looming biodiversity crisis would, from a conservationist perspective, appear to be dangerous things.

4.13   The Unnaturalness Argument Against De-extinction

The argument. De-extinction is unnatural and therefore wrong.

Commentary. People commonly object to the various methods and applications of synthetic biology—including *in vitro* fertilization, cloning, genetic engineering and de-extinction—on the basis of their 'unnaturalness', but it is notoriously unclear how such arguments are to be unpacked (Chadwick, 1982; Chapman, 2005; Cottrell et al., 2014; Norman, 1996; Sheehan, 2009; Takala, 2004; Van Den Belt, 2009). De-extinction would obviously be unnatural in the sense of being *the product of human agency,* but so is everything we do (Norman, 1996, p. 2; Rolston, 1979, p. 11), so this can't be the sense of word intended.

1. Is de-extinction unnatural in the sense that it produces an outcome (namely, the resurrection of an extinct species) that is novel and could not have arisen except through human technological intervention (Mason, 2017, p. 43)? Yes, but innumerable outcomes that are incontrovertibly *good* (at least from an environmentalist perspective) are also unnatural in this sense of the term (e.g., the replacement of dirty coal power with clean solar power, or the technology-enabled eradication of an invasive species from an offshore island). *Technology-enabled* outcomes are not necessarily *bad* outcomes, and to assume otherwise is to succumb to a version of the naturalistic fallacy.

2. Is de-extinction unnatural in the sense that it involves 'tampering with things we do not understand' in a way that is reckless, risky and hubristic (Chapman, 2005)? Maybe, but if this is the intended sense of the word 'unnatural' then the *unnaturalness argument* and the *hubris argument* are the same argument. We refer the reader back to our commentary on the hubris argument, above.

3. Is de-extinction unnatural because it produces inauthentic, synthetic, artificial organisms that will, if they are released into the natural world, undermine the integrity of the wilderness (Mason, 2017, pp. 48, 51–52; Sandler, 2013, p. 357)? No—at least not according to the authenticist position that was developed and defended in the previous chapter. To the contrary, de-extinct organisms will (at least potentially) be authentic to a high degree.

4. Is de-extinction unnatural in the sense that it "disturbs the course of nature" (Norman, 1996, p. 2)? Yes, but *all* conservation interventions do this, putting de-extinction in good company.

5. Finally, is de-extinction unnatural in the sense that it threatens a homeostatic 'balance of nature' (Rolston, 1979, pp. 14–16)? No, because as mentioned above (in our discussion of the balance restoration argument for de-extinction), ecologists deny that there is any such thing as a balance of nature. (Moreover, if there was a balance of nature then de-extinction would presumably tend to restore it rather that undermine it.)

4.14   The Animal Welfare Argument Against De-extinction

The argument. De-extinction should not be supported by conservationists because the de-extinction process will involve inflicting pain and suffering on animals, performing experiments on them, and keeping them in captivity (Kasperbauer, 2017).

Commentary. The animal welfare issue is moot for extinct species that aren't sentient, such as plants and invertebrates, but since they are seldom mentioned as de-extinction candidates (Turner 2017) this has little practical significance for the scope of the argument.

The main weakness in the animal welfare argument is its assumption that the goals of conservationism necessarily align with those of the animal welfare movement. This assumption is false. The goal of preserving and restoring biodiversity often conflicts with the goal of preventing animal suffering. One example, already mentioned above, is that of wildlife conservation in New Zealand, where umpteen millions of highly sentient mammals (possums, rats, cats, mustelids, goats and deer) are shot, trapped and poisoned on a yearly basis for the benefit of endemic and endangered plant, insect and bird species. (See Rolston (1985, p. 722) for a similar example.) Other examples are easy to find given how much suffering takes place in the natural ecosystems that conservationists seek to protect and restore. Consider these words of McMahan, Mill, Nussbaum, and Dawkins:

> Viewed from a distance, the natural world often presents a vista of sublime, majestic placidity. Yet beneath the foliage and hidden from the distant eye, a vast, unceasing slaughter rages. Wherever there is animal life, predators are stalking, chasing, capturing, killing, and devouring their prey. Agonized suffering and violent death are ubiquitous and continuous. (McMahan, 2010a)

> In sober truth, nearly all the things which men are hanged or imprisoned for doing to one another, are nature's everyday performances. (Mill & O'Grady, 1963, pp. 385–386)

> [T]he death of a gazelle after painful torture is just as bad for the gazelle when torture is inflicted by a tiger as when it is done by a human being … we have similar reasons to prevent it, if we can do so without doing greater harms. (Nussbaum, 2006, p. 379)

> The total amount of suffering per year in the natural world is beyond all decent contemplation. During the minute that it takes me to compose this sentence, thousands of animals are being eaten alive, many others are running for their lives, whimpering with fear, others are slowly being devoured from within by rasping parasites, thousands of all kinds are dying of starvation, thirst, and disease. It must be so. If there ever is a time of plenty, this very fact will automatically lead to an increase in the population until the natural state of starvation and misery is restored. (Dawkins, 1995)

If we had it as our aim to minimize animal suffering, rather than to conserve biodiversity, then we would be policing nature, by actively exterminating, or at least refusing to save, certain species. These include certain parasitic species (Naess, 1991) and predator species (Cowen, 2003; McMahan, 2010a, 2010b) that inflict great suffering on their hosts or prey. They also include certain sentient 'r-selected' prey species (with high fecundity) that are especially likely to suffer a painful and untimely death. From a purely animal-welfare-based perspective, their extinctions would, in all likelihood, be a net blessing.

In short, if de-extinction is proscribed by animal welfare considerations, then it is in good company because so too is a great deal of other wildlife

ETHICAL ARGUMENTS FOR AND AGAINST DE-EXTINCTION   115

conservation work. Conservationists are not playing on the same team as animal-welfarists. The two parties have different, incommensurable, and sometimes opposing objectives. Conservationists will typically give some weight to animal welfare considerations, but not at the expense of saving and restoring biodiversity.

4.15   THE TEACHABLE MOMENT ARGUMENT AGAINST DE-EXTINCTION

The argument. We should keep extinct species extinct because there are important lessons of humility to be learned by meditating upon their extinction. In Ben Minteer's words:

> [There] is great virtue in keeping extinct species extinct. Meditation on their loss reminds us of our fallibility and our finitude. We are a wickedly smart species, and occasionally a heroic and even exceptional one. But we are a species that often becomes mesmerized by its own power. It would be silly to deny the reality of that power. But we should also cherish and protect the capacity of nature, including those parts of nature that are no longer with us, to teach us something profound about the value of collective self-restraint and human limits. Few things teach us this sort of earthly modesty any more. (Minteer, 2014, p. 261)

> [In] fomenting the fantasy that we can erase the environmental abuses of the past by pursuing high-tech species revival technologies, promoters of de-extinction are inadvertently undermining the responsibility to learn the lessons of our environmental history. (Minteer, 2015, p. 15)

Commentary. One weakness in this argument is that for every extinct species we might successfully resurrect there will be innumerable others which—for want of DNA, or suitable surrogate species, or suitable habitat—are gone for good. Hence there is no fear of de-extinction ever robbing us of still-extinct species to teach us earthly modesty and remind us of our fallibility. A second weakness is that if lessons of humility are wanted, then they are surely better obtained by meditating on the many human-caused extinctions *that we will never be able to reverse* than on extinctions *that we could reverse but have chosen not to out of a conscious desire to teach ourselves a lesson of humility.* And, third, people seldom grieve too keenly for the loss of something they have never seen. Out of sight, out of mind.

To miss something, one must know what one is missing. Someone is much more likely to appreciate the value of protecting and restoring lost biodiversity by seeing a lost organism brought back to glorious life again than by reading about its extinction in a history book.

4.16   Conclusion: The Do's and Don'ts of Ethical De-extinction

In this chapter we have examined a raft of arguments for and against de-extinction within a conservationist context. The result is, as we have seen, a mixed bag. Of the arguments for de-extinction, most are weak, but some are forceful. The same goes for arguments against de-extinction: most are weak; some are forceful. Where does this leave us?

In a complicated place!

This should be no surprise. The technologies of de-extinction are not intrinsically good or bad. They are simply tools—extraordinarily powerful tools—that, like any tools, can be used either wisely or stupidly. De-extinction is neither good nor bad in and of itself. Whether it is good or bad depends instead on the details of the particular case in question, and on how the technology is applied. If it is used wisely and carefully, for the betterment of the environment, and for the betterment of the lives of future people, then it is good. If it is used foolishly or carelessly, then it is bad.

We will end by attempting to distil the lessons of the chapter into a set of 'do's and don'ts' for de-extinctions.

On timing

- Do 'freeze now and resurrect later'. Deferred de-extinctions are the order of the day.
- Do lay the groundwork for future de-extinctions by cryobanking genetically diverse cell lines from endangered species, and by documenting as much ecological information as possible about these species.
- Do promptly sequence or cryopreserve DNA from museum specimens of extinct species (because DNA degradation is inexorable at room temperature).

- Don't, except in exceptional circumstances, do de-extinction yet (on pain of falling foul of the resource allocation argument and the moral hazard argument).

On resourcing

- Do attract funding and talent that wouldn't otherwise go to conservation.
- Don't draw any funding or talent away from traditional conservation work (on pain of falling foul of the resource allocation argument).
- Don't initiate a de-extinction project unless there is funding, secured from outside the usual conservationist 'pot', to see the project through all of its stage, including captive rearing, wild release, and post-release management.

On candidate selection

- Do select a species for which there is suitable wild habitat, and where this habitat is both free of the causes of the original extinction and relatively secure against future threats like climate change, sea level rise, and ocean acidification.
- Do select a species that, for animal husbandry reasons, is amenable to captive rearing and translocation.
- Do select a species for which abundant, genetically diverse DNA sources exist, so that a genetically healthy de-extinct population can be created.
- Do favour species that made an outsized contribution to genetic diversity (because of their ecological importance as keystone species, or because of their phenotypic or phylogenetic uniqueness), or that were of especial cultural significance to a people.
- Do favour species that had a fast breeding cycle and a high fertility rate (so that a large population can be quickly re-established).
- Do favour species that had a high effective population size (i.e., species in which most organisms of both sexes reproduce) in order to minimize genetic drift, inbreeding depression, and loss of evolutionary potential (Steeves et al., 2017).

- Don't select a species that presents a high risk of becoming an 'invasive from the past' or a vector for disease transmission.
- Do select a species the wild-release of which is reversible in case something goes wrong.
- Don't select a species the rewilding of which would be politically infeasible due to the opposition of stakeholders.
- Do select a species about which enough is known to accurately assess whether it meets the above criteria.

On publicity

- Do use the media platform that de-extinction provides to promote and support traditional conservation work, especially habitat protection.
- Do integrate the project with traditional conservation projects, so that they benefit from public exposure too.
- Do minimize the moral hazard associated with the project, by making it abundantly clear to the public that de-extinction is an uncertain method of last resort, which can be applied only to species satisfying a very restrictive list of criteria.
- Don't hype de-extinction, or oversell what it can achieve, and don't on any account paint it as being a silver bullet that means we don't need to take the biodiversity crisis seriously anymore.

On methodology

- Do consider performing a discriminating de-extinction (in order to boost the initial reproductive vigour of the de-extinct population).
- Do follow the usual ethical protocols for eliminating or minimizing animal suffering.
- Do follow established guidelines for captive rearing (McGowan, Traylor-Holzer, & Leus, 2017) and translocation (IUCN/SSC, 2013)—and de-extinction-specific guidelines based on these (IUCN/SSC, 2016).

Projects that satisfy all the above do's, and none of the above don'ts, should, in our view, be supported, not opposed, by conservationists.

NOTES

1. "The gene is the basic unit of selfishness" (Dawkins, 1976, p. 39). This being so, could we not identify a species with its genepool, and in this way substantiate the claim that a species 'wants' to remain extant and perpetuate itself into the future? No, for in the first place what is in the selfish interests of an individual gene is often sharply detrimental to the interests of other genes and to long-term survival prospects of the genepool as a whole. The interests of the genes therefore don't add up to any coherent volition at the genepool level. In the second place, attributing selfishness to genes is just a useful figure of speech, that involves anthropomorphising a segment of DNA. Genes are 'selfish' and have 'interests' only in the extremely deflationary sense that self-perpetuating chain letters are selfish and have interests.

2. See Jebari (2016) and Cohen (2014) for attempts to show that we owe a moral duty to species we have exterminated, a duty that requires us to resurrect them. Cohen acknowledges the profound difficulties confronting this view, writing: "The potential objections to [my] analysis ... are legion. My aim was to show that the initially absurd-sounding idea of a duty of de-extinction deserves a second thought" (p. 172).

3. This assumption is not needed by proponents of the justice argument in the special case that *our generation* is responsible for the extinction, but in practice proponents of the justice argument are usually arguing for the resurrection of species that went extinct long ago.

4. The creation of de-extinct organisms for zoos or for the exotic pet industry would be of substantial conservation benefit only if some of the created organisms became available for rewilding (although see Archer (2013) for a contrary view). Depending on the species, it could also raise obvious animal welfare objections.

5. There are some time constraints. Ideally a species will be resurrected sooner rather than later, especially when fast-paced environmental or ecological changes would make rewilding the species rapidly more difficult the longer the delay (Robert, Thévenin, Princé, Sarrazin, & Clavel, 2017). Another reason to resurrect sooner rather than later is so that more generations of people can benefit from the species' de-extinction.

REFERENCES

Archer, M. A. (2013). *Second chance for Tasmanian tigers and fantastic frogs.* Presented at the TEDx DeExtinction/National Geographic, Washington, DC. Retrieved from http://longnow.org/revive/tedxdeextinction/

Benirschke, K. (1984). The frozen zoo concept. *Zoo Biology, 3*(4), 325–328. https://doi.org/10.1002/zoo.1430030405

Bennett, J. R., Maloney, R. F., Steeves, T. E., Brazill-Boast, J., Possingham, H. P., & Seddon, P. J. (2017, March 1). Spending limited resources on de-extinction could lead to net biodiversity loss. *Nature Ecology & Evolution, 1*(4), 53.

Blockstein, D. E. (2017). We can't bring back the passenger pigeon: The ethics of deception around de-extinction. *Ethics, Policy & Environment, 20*(1), 33–37. https://doi.org/10.1080/21550085.2017.1291826

Botkin, D. B. (1992). *Discordant harmonies: A new ecology for the twenty-first century*. Oxford: Oxford University Press.

Brand, S. (2013). *The dawn of de-extinction: Are you ready?* Presented at the TED, Long Beach, CA. Retrieved from http://www.ted.com/talks/stewart_brand_the_dawn_of_de_extinction_are_you_ready.html

Brand, S. (2014). De-extinction debate: Should we bring back the woolly mammoth? *Yale Environment, 360*. Retrieved from http://e360.yale.edu/features/the_case_for_de-extinction_why_we_should_bring_back_the_woolly_mammoth

Brown, S. (2016, July 28). Heath hen tops de-extinction list. Vineyard Gazette.

Burney, D. A., Juvik, J. O., Burney, L. P., & Diagne, T. (2012). Can unwanted suburban tortoises rescue native Hawaiian plants? *The Tortoise, 104–115.*

Callaway, E. (2016). Geneticists aim to save rare rhino. *Nature, 533*, 20–21.

Campbell, D. I. (2016). A case for resurrecting lost species—Review essay of Beth Shapiro's, "How to clone a mammoth: The science of de-extinction". *Biology and Philosophy, 31*(5), 747–759. https://doi.org/10.1007/s10539-016-9534-2

Carlin, N. F., Wurman, I., & Zakim, T. (2013). How to permit your mammoth: Some legal implications of de-extinction. *Stanford Environmental Law Journal, 33*(1), 3–57.

Caro, T. (2010). *Conservation by proxy: Indicator, umbrella, keystone, flagship, and other surrogate species*. Washington, DC: Island Press.

Chadwick, R. F. (1982). Cloning. *Philosophy, 57*(220), 201–209. https://doi.org/10.1017/S0031819100050774

Chapman, A. (2005). Genetic engineering: The unnatural argument. *Techné: Research in Philosophy and Technology, 9*(2), 81–93.

Cohen, S. (2014). The ethics of de-extinction. *NanoEthics, 8*(2), 165–178.

Cottrell, S., Jensen, J. L., & Peck, S. L. (2014). Resuscitation and resurrection: The ethics of cloning cheetahs, mammoths, and Neanderthals. *Life Sciences, Society and Policy, 10*(1), 3. https://doi.org/10.1186/2195-7819-10-3

Cowen, T. (2003). Policing nature. *Environmental Ethics, 25*(2), 169–182.

Crist, E. (2008). Cloning in restorative perspective. In M. Hall (Ed.), *Restoration and history: The search for a usable environmental past* (pp. 284–292). New Brunswick, NJ: Rutgers University Press.

Dawkins, R. (1976). *The selfish gene*. Oxford: Oxford University Press.

Dawkins, R. (1995). *River out of Eden: A Darwinian view of life*. London: Weidenfeld & Nicolson.

Diehm, C. (2017). De-extinction and deep questions about species conservation. *Ethics, Policy & Environment*, *20*(1), 25–28. https://doi.org/10.1080/2155 0085.2017.1291827

Donlan, C. J. (2014). De-extinction in a crisis discipline. *Frontiers of Biogeography*, *6*(1), 25–28.

Donlan, C. J., Berger, J., Bock, C. E., Bock, J. H., Burney, D. A., Estes, J. A., … Greene, H. W. (2006). Pleistocene rewilding: An optimistic agenda for twenty-first century conservation. *The American Naturalist*, *168*(5), 660–681. https://doi.org/10.1086/508027

Ehrlich, P., & Ehrlich, A. (2014). The case against de-extinction: It's a fascinating but dumb idea. *Yale Environment*, 360. Retrieved from http://e360.yale.edu/features/the_case_against_de-extinction_its_a_fascinating_but_dumb_idea

Hansford, D. (2016). *Protecting paradise: 1080 and the fight to save New Zealand's wildlife*. Nelson, NZ: Potton & Burton.

Iacona, G., Maloney, R. F., Chadès, I., Bennett, J. R., Seddon, P. J., & Possingham, H. P. (2017). Prioritizing revived species: What are the conservation management implications of de-extinction? *Functional Ecology*, *31*(5), 1041–1048. https://doi.org/10.1111/1365-2435.12720

IUCN. (2017). The IUCN red list of threatened species. Version 2017-1. IUCN. Retrieved from www.iucnredlist.org

IUCN/SSC. (2013). Guidelines for reintroductions and other conservation translocations (Version 1.0. Gland). Switzerland: IUCN Species Survival Commission.

IUCN/SSC. (2016). Guiding principles on creating proxies of extinct species (Version 1.0. Gland). Switzerland: IUCN Species Survival Commission.

Jebari, K. (2016). Should extinction be forever? *Philosophy and Technology*, *29*(3), 211–222.

Jones, K. E. (2014). From dinosaurs to dodos: Who could and should we de-extinct? *Frontiers of Biogeography*, *6*(1), 20–24.

Jørgensen, D. (2013). Reintroduction and de-extinction. *Bioscience*, *63*(9), 719–720. https://doi.org/10.1525/bio.2013.63.9.6

Kasperbauer, T. J. (2017). Should we bring back the passenger pigeon? The ethics of de-extinction. *Ethics, Policy & Environment*, *20*(1), 1–14. https://doi.org/10.1080/21550085.2017.1291831

Kohl, P. (2017). Using de-extinction to create extinct species proxies; Natural history not included. *Ethics, Policy & Environment*, *20*(1), 15–17. https://doi.org/10.1080/21550085.2017.1291832

Mason, C. (2017). The unnaturalness objection to de-extinction: A critical evaluation. *Animal Studies Journal*, *6*(1), 40–60.

McCarthy, D. P., Donald, P. F., Scharlemann, J. P. W., Buchanan, G. M., Balmford, A., Green, J. M. H., … Butchart, S. H. M. (2012). Financial costs of meeting global biodiversity conservation targets: Current spending and unmet needs. *Science, 338*(6109), 946–949. https://doi.org/10.1126/science.1229803

McCauley, D. J., Hardesty-Moore, M., Halpern, B. S., & Young, H. S. (2017). A mammoth undertaking: Harnessing insight from functional ecology to shape de-extinction priority setting. *Functional Ecology, 31*(5), 1003–1011. https://doi.org/10.1111/1365-2435.12728

McGowan, P. J. K., Traylor-Holzer, K., & Leus, K. (2017). IUCN guidelines for determining when and how ex situ management should be used in species conservation. *Conservation Letters, 10*(3), 361–366. https://doi.org/10.1111/conl.12285

McMahan, J. (2010a, September 19). The meat eaters. *The New York Times.*

McMahan, J. (2010b, September 28). Predators: A response. *The New York Times.*

Mill, J. S., & O'Grady, J. (1963). *Collected works of John Stuart Mill.* Toronto: University of Toronto Press.

Minteer, B. A. (2014). Is it right to reverse extinction? *Nature, 509*(7500), 261.

Minteer, B. A. (2015). The perils of de-extinction. *Minding Nature, 8*(1), 11–17.

Naess, A. (1991). Should we try to relieve clear cases of suffering in nature? *Pan Ecology, 6,* 1–5.

Nes, E. H. van, & Scheffer, M. (2004). Large species shifts triggered by small forces. *The American Naturalist, 164*(2), 255–266. https://doi.org/10.1086/422204

Norman, R. (1996). Interfering with nature. *Journal of Applied Philosophy, 13*(1), 1–12. https://doi.org/10.1111/j.1468-5930.1996.tb00144.x

Nussbaum, M. (2006). *Frontiers of justice: Disability, nationality, species membership.* Cambridge, MA: Belknap Press.

O'Connor, M. R. (2015). *Resurrection science: Conservation, de-extinction and the precarious future of wild things.* New York: St. Martin's Press.

Pickett, S. T. A. (2013). The flux of nature: Changing worldviews and inclusive concepts. In R. Rozzi, S. T. A. Pickett, C. Palmer, J. J. Armesto, & J. B. Callicott (Eds.), *Linking ecology and ethics for a changing world: Values, philosophy, and action* (pp. 265–279). Dordrecht: Springer Netherlands. https://doi.org/10.1007/978-94-007-7470-4_23

Pimm, S. (2013). The case against species revival. *National Geographic.* Retrieved from http://news.nationalgeographic.com/news/2013/03/130312--deextinction-conservation-animals-science-extinction-biodiversity-habitat-environment/

Redford, K. H., Adams, W., & Mace, G. M. (2013). Synthetic biology and conservation of nature: Wicked problems and wicked solutions. *PLoS Biology, 11*(4), 1–4. https://doi.org/10.1371/journal.pbio.1001530

Redford, K. H., & Sanjayan, M. A. (2003). Retiring cassandra. *Conservation Biology, 17*(6), 1473–1474.

Robert, A., Thévenin, C., Princé, K., Sarrazin, F., & Clavel, J. (2017). De-extinction and evolution. *Functional Ecology*, *31*(5), 1021–1031. https://doi.org/10.1111/1365-2435.12723

Rolston, H. (1979). Can and ought we to follow nature? *Environmental Ethics*, *1*(1), 7–30.

Rolston, H. (1985). Duties to endangered species. *Bioscience*, *35*(11), 718–726.

Ryder, O. A., McLaren, A., Brenner, S., Zhang, Y.-P., & Benirschke, K. (2000). DNA banks for endangered animal species. *Science*, *288*(5464), 275–277. https://doi.org/10.1126/science.288.5464.275

Sandler, R. (2013). The ethics of reviving long extinct species. *Conservation Biology*, *28*(2), 354–360.

Sandler, R. (2017). De-extinction: Costs, benefits and ethics. *Nature Ecology & Evolution*, *1*, 105.

Scientific American Editors. (2013, May 14). Why efforts to bring extinct species back from the dead miss the point. *Scientific American*. Retrieved from https://www.scientificamerican.com/article/why-efforts-bring-extinct-species-back-from-dead-miss-point/

Seddon, P. J. (2017). The ecology of de-extinction. *Functional Ecology*, *31*(5), 992–995. https://doi.org/10.1111/1365-2435.12856

Seddon, P. J., Moehrenschlager, A., & Ewen, J. (2014). Reintroducing resurrected species: Selecting deextinction candidates. *Trends in Ecology & Evolution*, *29*(3), 140–147. https://doi.org/10.1016/j.tree.2014.01.007

Sheehan, M. (2009). Making sense of the immorality of unnaturalness. *Cambridge Quarterly of Healthcare Ethics*, *18*(2), 177–188. https://doi.org/10.1017/S096318010909029X

Sherkow, J. S., & Greely, H. T. (2013). What if extinction is not forever? *Science*, *340*(6128), 32–33. https://doi.org/10.1126/science.1236965

Simberloff, D. (2014). The "balance of nature"—Evolution of a panchreston. *PLoS Biology*, *12*(10), 1–4. https://doi.org/10.1371/journal.pbio.1001963

Snyder, N. F. R., Derrickson, S. R., Beissinger, S. R., Wiley, J. W., Smith, T. B., Toone, W. D., & Miller, B. (1996). Limitations of captive breeding in endangered species recovery. *Conservation Biology*, *10*(2), 338–348. https://doi.org/10.1046/j.1523-1739.1996.10020338.x

Steeves, T. E., Johnson, J. A., & Hale, M. L. (2017). Maximising evolutionary potential in functional proxies for extinct species: A conservation genetic perspective on de-extinction. *Functional Ecology*, *31*(5), 1032–1040. https://doi.org/10.1111/1365-2435.12843

Swart, S. (2015). Zombie zoology: History and reanimating extinct animals. In S. Nance (ed.), *The historical animal* (pp. 54–72). Syracuse, NY: Syracuse University Press. Retrieved from http://www.jstor.org/stable/j.cttlj2n79q.7

Takala, T. (2004). The (im)morality of (un)naturalness. *Cambridge Quarterly of Healthcare Ethics*, *13*(1), 15–19. https://doi.org/10.1017/S0963180104131046

Turner, D. D. (2017). Biases in the selection of candidate species for de-extinction. *Ethics, Policy & Environment, 20*(1), 21–24. https://doi.org/10.1080/21550085.2017.1291835

Van Den Belt, H. (2009). Playing God in Frankenstein's footsteps: Synthetic biology and the meaning of life. *NanoEthics, 3.* https://doi.org/10.1007/s11569-009-0079-6

Waldron, A., Mooers, A. O., Miller, D. C., Nibbelink, N., Redding, D., Kuhn, T. S., ... Gittleman, J. L. (2013). Targeting global conservation funding to limit immediate biodiversity declines. *Proceedings of the National Academy of Sciences, 110*(29), 12144–12148. https://doi.org/10.1073/pnas.1221370110

Welchman, J. (2017). How much is that mammoth in the window? *Ethics, Policy & Environment, 20*(1), 41–43. https://doi.org/10.1080/21550085.2017.1299674

Whittle, P. M., Stewart, E. J., & Fisher, D. (2015). Re-creation tourism: De-extinction and its implications for nature-based recreation. *Current Issues in Tourism, 18*(10), 908–912. https://doi.org/10.1080/13683500.2015.1031727

Wu, J., & Loucks, O. L. (1995). From balance of nature to hierarchical patch dynamics: A paradigm shift in ecology. *The Quarterly Review of Biology, 70*(4), 439–466. https://doi.org/10.1086/419172

Zorich, Z. (2010). Should we clone Neanderthals? The scientific, legal, and ethical obstacles. *Archaeology, 63.*

Index[1]

[1] Notes: Page numbers followed by 'n' refers to notes.

© The Author(s) 2017
D.I. Campbell, P.M. Whittle, *Resurrecting Extinct Species*,
https://doi.org/10.1007/978-3-319-69578-5